DISCOVERING THE UNIVERSE

DISCOVERING THE UNIVERSE

A guide to the galaxies, planets, and stars

Sten Odenwald

Dedication

In remembrance of the many victims of the Covid-19 pandemic whose curiosity about the world around them lives on.

Other books by Sten Odenwald

The 23rd Cycle: Learning to Live with a Stormy Star
Patterns in the Void: Why Nothing is Important
The Astronomy Café: 365 Questions and Answers from Ask the Astronomer
Back to the Astronomy Café
My Astronomical Life: A First-person Journey
Cosmic History I: From the Big Bang to the last Ice Age
Cosmic History II: From the Ice Age to the End of Time
Eternity: A User's Guide
Interstellar Travel: An Astronomer's Guide
Solar Storms: 2000 years of Human Calamity!
Exploring Quantum Space
A Degree in a Book: Cosmology
Knowledge in a Nutshell: Astrophysics
Space Exploration: A History in 100 objects
How the World Works: Quantum Physics

ARCTURUS
This edition published in 2021 by Arcturus Publishing Limited
26/27 Bickels Yard, 151–153 Bermondsey Street,
London SE1 3HA

ISBN: 978-1-83940-869-4
AD008397US

Printed in Singapore

Contents

NASA

// Introduction

Ever since humans appeared, we have been on a journey to discover the world around us. It was originally a matter of survival to be aware of the landscape, the regularity of the seasons, and the nature and locations of predators and prey. Over recent centuries, this process of discovery has been compelled less by any survival imperative and more by sheer curiosity. The technological fruits of this curiosity have utterly transformed our civilization, especially during the 20th century. Among the overarching questions whose answers are pursued by this curiosity are the contents, structure, and nature of our universe: How were the sun and Earth formed? What is the origin and destiny of our universe? Are we alone in the cosmos?

For millennia, philosophers attempted to answer these questions, but failed to make any lasting progress. You cannot answer these kinds of questions through semantic manipulations or deductive logic. You need raw information that the ordinary senses are incapable of providing. It wasn't until the advent of the telescope in the 1600s and the spectroscope in the 1800s that scientists acquired the technology to dramatically extend the senses and gather crucial data about the sun, planets, and stars.

Sir Isaac Newton once said that his work was the result of "standing on the shoulders of giants." No less is true of where we find ourselves today. It has taken generations of scientists and millions of hours of labor to reach a point in human history where ancient questions could at last find their answers. We have discovered our universe, not as a mysterious and inscrutable abstraction but as a concrete and knowable system of matter, energy, space, and time. At the same time, it is filled with wondrous and amazing objects and events, not the least of which is our own origins as sentient beings allowing the universe to comprehend itself.

Below *A photograph by the Hubble Space Telescope depicting "Mystic Mountain," a pillar of gas and dust three light-years high in the Carina Nebula.*

HOW TO BUILD A UNIVERSE

The term universum was first coined by the Roman statesman Cicero in the 1st century BCE. Today, we know that our universe includes all things on Earth, our solar system, and the distant stars and galaxies beyond. It also encompasses a vast and possibly infinite space, which has been in existence for nearly 14 billion years. The manner in which the universe came into existence was for most of human history a matter of religious consideration. All of the creation stories shared one thing in common: they had to provide an explanation for how something (the universe) was created or appeared out of nothing. Today, astrophysicists are still struggling with this perplexing mystery expressed in modern language, and with a modicum of impish humour, as "Why is there Something rather than Nothing?"

Hubble eXtreme Deep Field (XDF) image of a small portion of the universe showing thousands of galaxies to a distance of nearly 13 billion light years. At its farthest limits it can just detect infant galaxies formed 500 million years after the Big Bang.

// Creation stories

In antiquity, the composition of the objects in the universe was based upon a set of basic elements proposed by Aristotle as earth, air, fire, water, and aether. The first four were found on Earth while the planets, stars, and other denizens of the *Empyrium* were fashioned from a pure, luminous substance called aether (αιθερ). Although Aristotle considered five elements, in India's Vedic philosophy these were supplemented by the elements of time, direction, mind and soul. All objects in the world were fashioned from combinations of these elements. Ancient philosophers who thought about the universe invariably found themselves thinking in terms of the basic ingredients to all things, which came to be called *atomos* by the 5th century BCE Greeks such as Democritus, or *parmanu* in the 6th century BCE by the Vedic sage, Kanada.

Alongside a knowledge of the ingredients to the world, people had to create stories to explain how specific things in the world came to be formed from these elemental ingredients. In ancient Egypt, Atum-Ra first created himself out of the dark waters of Nun by uttering his own name. He followed this act by bringing into being over time all the other gods and places. Babylonians also began with cosmic waters imbued with their own deities: Apsu for fresh water and Tiamat for bitter salt water. These conflicting deities then created all the other deities including Marduk who eventually kills Tiamat, and from her corpse creates the heavens and Earth. Also in the mid-East, the Judeo-Christian Genesis of the Old Testament begins with the formless waters of Tehom that were acted upon by Elohim (Yahweh) to create the heavens, Earth, and all life.

The biggest challenge for our ancestors in fashioning these stories is stated perhaps for the first time in the *Rig Veda* (10:129): "Who knows from whence this great creation sprang?" It is answered by the realization that even the "most-High seer that is in highest heaven" may not know! Nevertheless, for thousands of years, humans found these kinds of stories entirely workable and useful for their needs. Only in the last 100 years have new insights allowed us to fashion an even better "story" of what we now call cosmogenesis. The biggest challenge for humans today has been in incorporating into the modern Creation "story" all of the new, essential ingredients we have uncovered and also showing how they are interrelated in a logical way. These ingredients represent seven specific kinds of phenomena and attributes of our physical world. Let's have a look at them one by one.

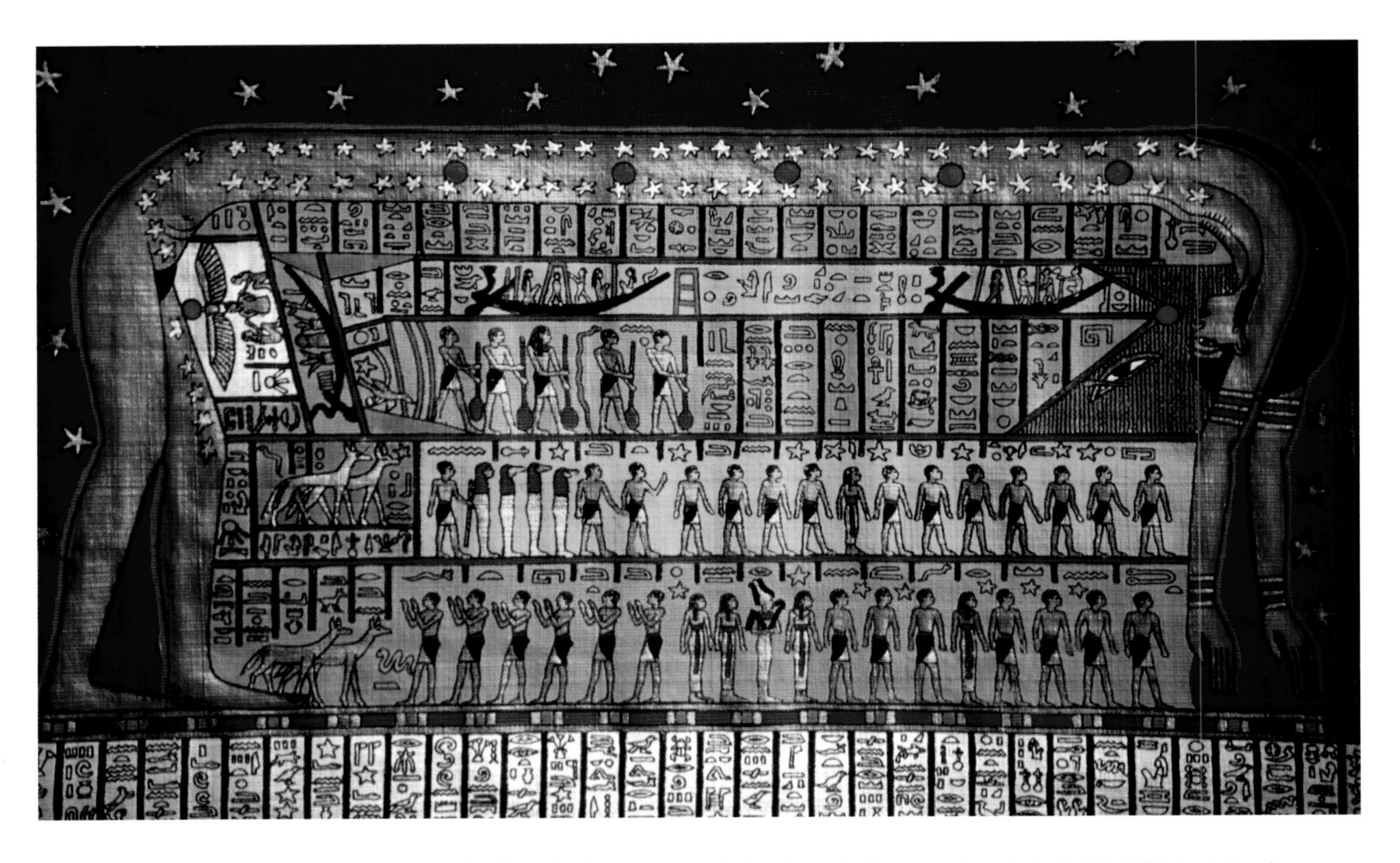

Above *Egyptian universe after Atum-Ra creates Shu (Air), who then separates the heavens (Nut) from the land (Geb).*

Below Chaos *by George Frederic Watts (1817–1904) depicting the primeval state of nature described in the Biblical Book of Genesis.*

// Ingredient 1: Matter

By as late as the 15th century alchemists had only succeeded in identifying a few dozen additional compounds beyond Aristotle's canonical five. Today, the search for the fundamental elements of nature has inexorably led to the discovery of more than 94 naturally occurring ones on Earth, and an additional 24 artificially created with advanced technology. Centuries of scientific investigation and technological advance have also led to a deep understanding of the nature of matter formed from a small collection of basic elements assembled into molecules of bewildering complexity. But the reduction of matter to its most elementary constituents did not end here. Starting in the early 20th century, atoms were also found to be composed of electrons, protons, and neutrons. The heaviest known element, called Oganesson, was discovered in 2002 and has 118 protons, 118 electrons, and 176 neutrons. By mid-century, experiments on protons discovered that they, themselves, were composed of still more elementary objects called quarks. Over time, physicists discovered exactly six different kinds of quarks that were given the humorous names: up (U), down (D), strange (S), charm (C), bottom (B), and top (T). The familiar protons and neutrons only required two of these types, the U and D quarks, assembled into groups of three, for example, a proton consists of the three-quark combination UUD and the neutron has the opposite combination DDU. Hundreds of other, more massive, particles required additional combinations of the six types of quarks to account for them.

Alongside the six quarks, the second and much lighter family of elementary particles are called the leptons. The electron, a workhorse of our modern civilization, is the most familiar of these, but is paired with another particle called the neutrino. During the process of radioactive decay, such as when a neutron decays after about ten minutes, the neutron becomes a proton and also emits an electron and a neutrino. Other particles can also undergo decays emitting additional, more-massive leptons such as the muon and the tauon, accompanied by their own partnering neutrinos.

Antimatter was discovered in the 1930s and is a form of matter in which its charge has an opposite sign from normal quark and lepton matter. For example, the electron with a negative electric charge has an antimatter version called the positron with exactly the same mass but with a positive charge. A down quark (D) with a charge of $-\frac{1}{3}$ has an antimatter version (*D*) with a $+\frac{1}{3}$ electric charge. Similarly, the up quark (U) with $+\frac{2}{3}$ charge has an antimatter version

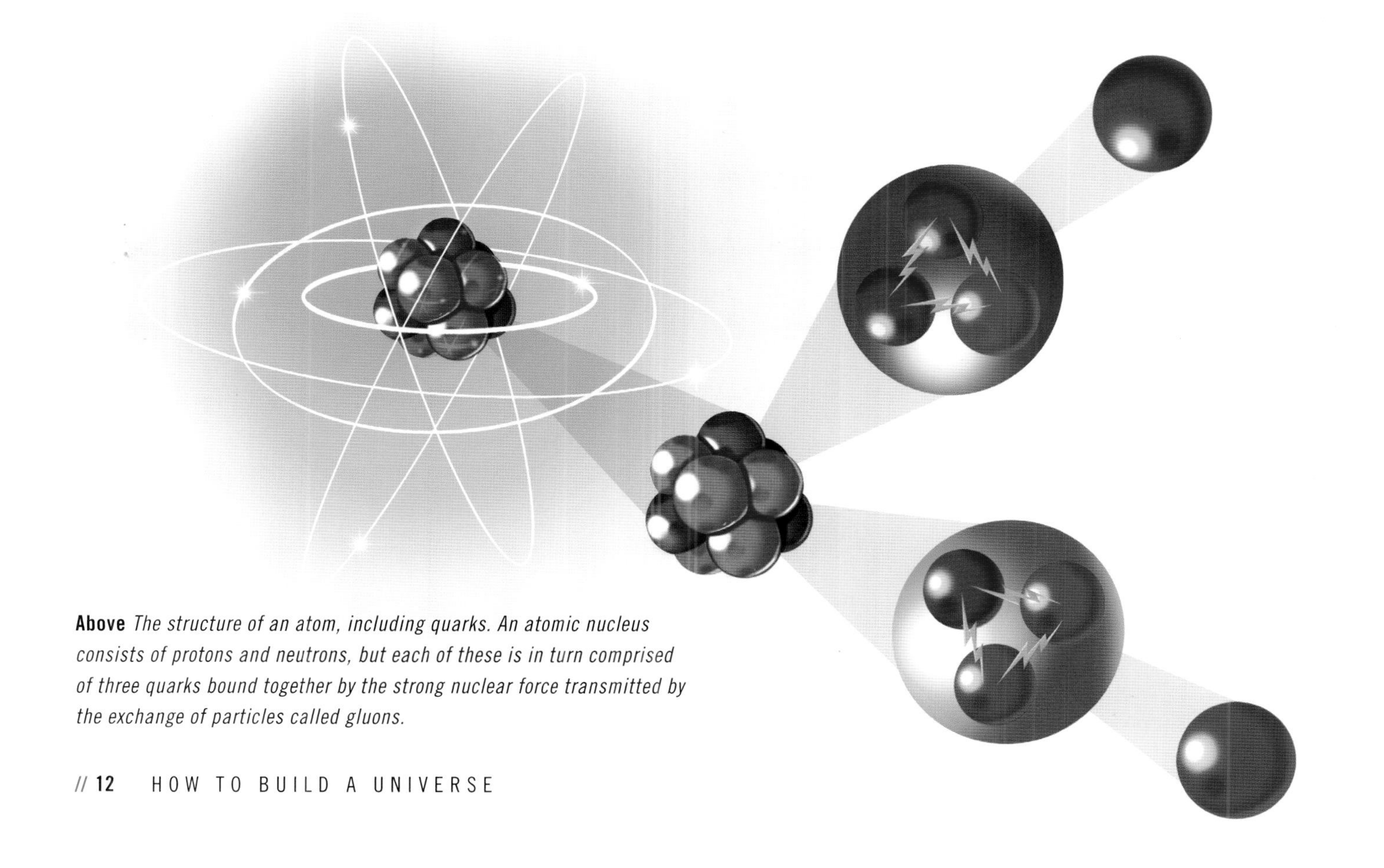

Above *The structure of an atom, including quarks. An atomic nucleus consists of protons and neutrons, but each of these is in turn comprised of three quarks bound together by the strong nuclear force transmitted by the exchange of particles called gluons.*

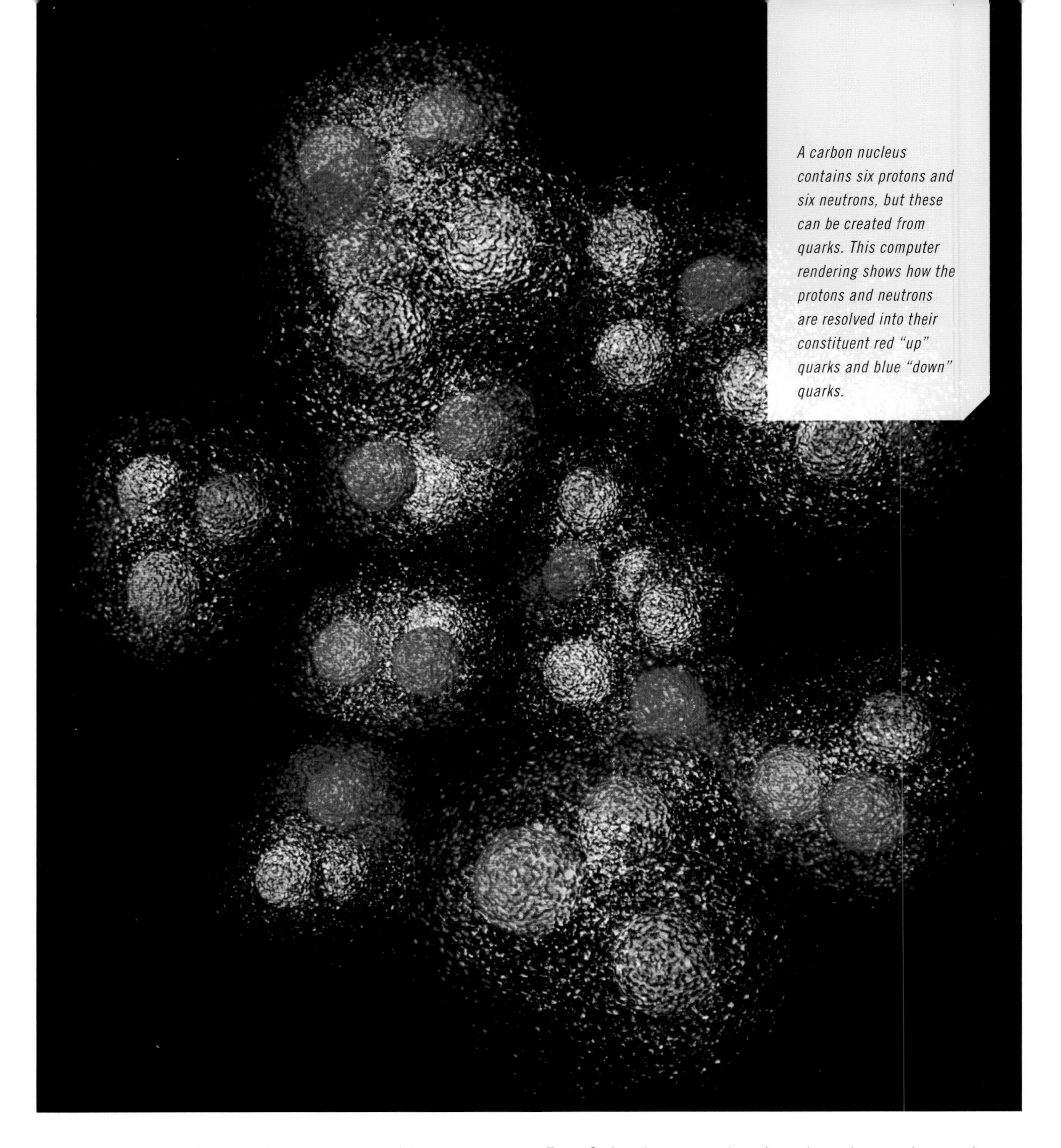

A carbon nucleus contains six protons and six neutrons, but these can be created from quarks. This computer rendering shows how the protons and neutrons are resolved into their constituent red "up" quarks and blue "down" quarks.

(*U*) with -⅔ charge. This is why the neutron with no net charge has an antiparticle, the antineutron. While the neutron contains the matter quarks DDU, the antineutron consists of the three antiquarks DDU.

Another important feature of particles and their antiparticles is that, when they are brought together, they vanish in a burst of energy. Albert Einstein's theory of special relativity states that matter and energy are equivalent physical properties, related by his iconic formula $E=mc^2$. An electron and positron brought together produce exactly two gamma rays that each carry off an amount of energy equal to $E=mc^2$ where m is the mass of an electron and c is the speed of light. It is also possible to create an electron-positron pair by using an "atom smasher" where the collision energy between particles can be used to create these pairs almost literally "out of nothing." The essential contents of our universe can be neatly summarized by the six quarks, the six leptons and their antimatter twins, and

are codified in what physicists call the Standard Model. But there is a fly in the ointment.

Since the 1990s, astronomers have studied the movements of galaxies and the rotation of our own Milky Way, uncovering a vast reservoir of unseen "dark" matter. Dark matter is not the same kind of matter that appears in the Standard Model. It appears to be invisible; it emits no light; nor does it seem to absorb or reflect light from more distant stars. Instead, it can only be detected by its gravitational influences on things we can see. The motions of galaxies near the Milky Way, as well as the speeds and movements of stars and gas clouds inside the Milky Way, reveal the extent of a vast halo of dark matter surrounding our own galaxy. Modern estimates suggest that about eight times more dark matter exists in our Milky Way than in the visible, ordinary matter that makes up the stars and gas clouds. This dominance of dark matter can also be detected in many nearby galaxies. Without substantial amounts of dark matter, many galaxies would simply spin apart instead of persisting as they seem to do for billions of years.

Although no new quarks or leptons have been found despite 50 years of effort, the search for dark matter remains one of the most exciting activities in contemporary astrophysics. Astrophysicists have tried to use the existence of neutrinos in the Standard Model as candidates. By imbuing neutrinos with 1⁄100,000 the mass of an electron, they could collectively have enough mass and gravity to mimic dark matter. This was an exciting prospect in the 1980s and 1990s until precise measurements of the mass of the three known types of neutrinos showed that they were insufficient to provide enough gravity. From purely theoretical considerations, physicists have identified candidate particles for dark matter by extending the Standard Model to higher energies using a new principle called supersymmetry. The most promising is called the neutralino. At the Large Hadron Collider (LHC) in Geneva, Switzerland, and despite a decade of search, no evidence for either supersymmetry or this new particle has been found. Between the mass of the heaviest "top" quark, at just over 180 times that of a single proton, to the limit of the LHC at 10,000 times the mass of a proton, no new particles have been detected. This particle desert is a startling and entirely unprecedented experience for physicists. Nature seems to have run out of new forms of matter beyond what we have already catalogued in the Standard Model.

Right *Location of dark matter in the cluster of galaxies known as Abell 1689. The spots are the individual galaxies and the white cloudiness is the estimated location of dark matter colorized to show its location.*

// Ingredient 2: The fundamental forces of nature

The next ingredients to our universe are the forces that make matter "do" something interesting. Without forces, the universe would be a static gas of quarks and leptons in space. Since ancient times, humans have known about the first of these elementary forces called electromagnetism. This is the force that allowed the magnetic compasses of ancient Chinese mariners to work, or to deliver electrical shocks when amber is rubbed with fur as discovered by the ancient Greeks.

Charged particles possess electric fields that radiate away from the objects like the spokes in a wheel. This field will produce a force on another charged object that it encounters, causing the familiar attraction or repulsion if the charges are opposite (attraction) or the same (repulsion). These fields-of-force are also responsible for the rigidity of rocks, mountains, planets, and even humans. When charged particles are in motion, they also produce magnetic fields, which we commonly see in such things

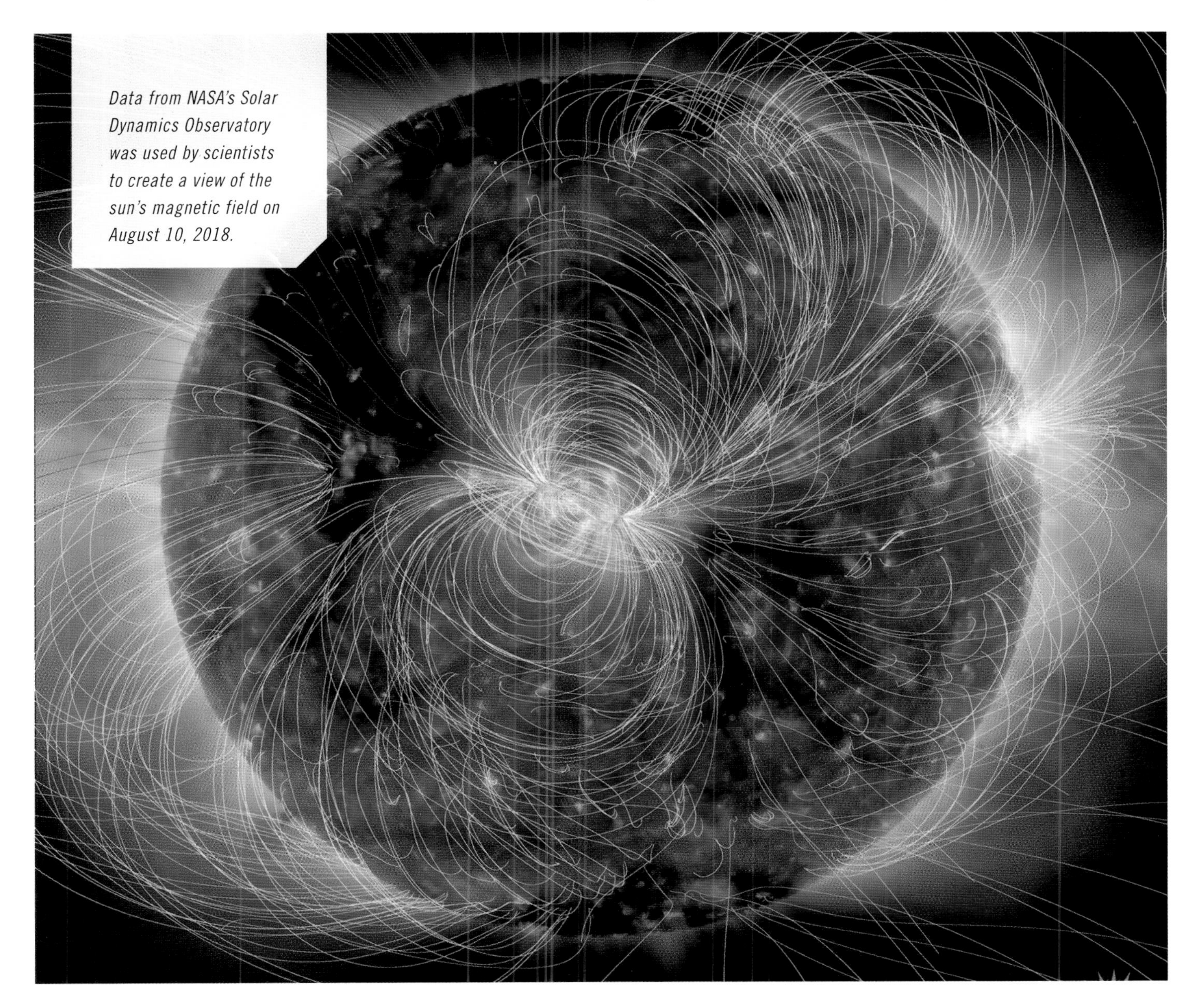

Data from NASA's Solar Dynamics Observatory was used by scientists to create a view of the sun's magnetic field on August 10, 2018.

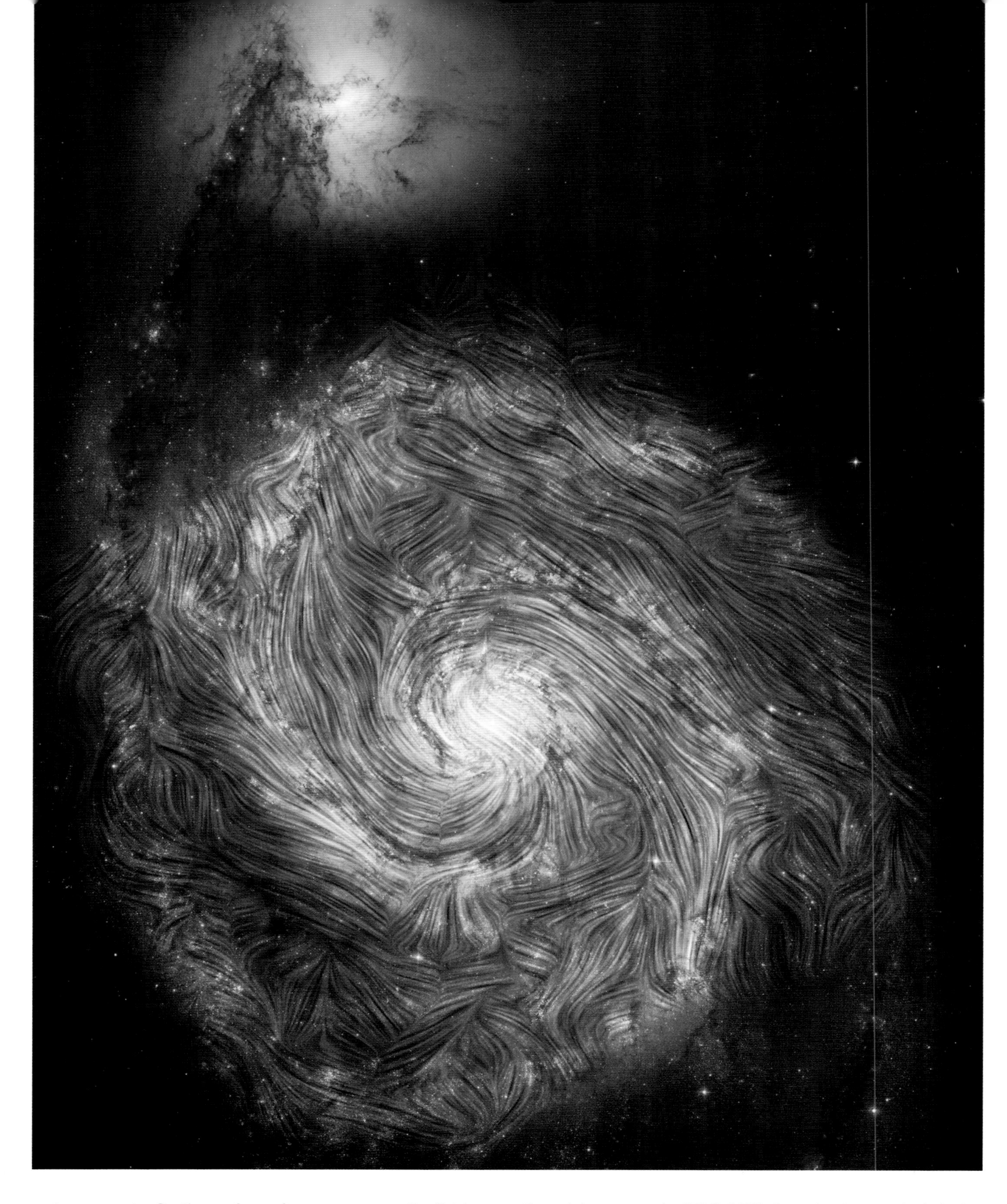

as toy magnets. On the surface of our sun, magnetic fields caused by the motion of electrically-charged gases called plasmas can become so strong they pop through the solar surface to become sunspots. Sunspots are born in pairs with one member having a north-type and the other a south-type

Above *Astronomers using NASA's SOFIA observatory have used polarized light to map out the magnetic fields in the nearby Whirlpool Galaxy (Messier 51). These fields are caused by flows in the ionized interstellar medium, much like electrical currents in an ordinary copper wire cause magnetic fields.*

Magnetic fields in the galaxy NGC 1086. The magnetic fields align along the entire length of the massive spiral arms—24,000 light-years across.

polarity just like a toy magnet. The motion of the plasma can drag these magnetic fields around, amplifying them and allowing them to affect even more distant regions of the sun.

Electrons in an atom are held together by long-range electromagnetic forces, but atomic nuclei would fly apart due to the intense electromagnetic repulsion from all of the positive nuclear protons. To bind the quarks into protons and neutrons, and keep these particles confined to atomic nuclei, a short-range and very strong force is required. This strong nuclear force is caused by the exchange of particles called gluons. Gluons resemble photons of light in that they carry no mass at all, but interact with all nuclear particles consisting of quarks. Unlike photons, which only come in one flavor, there are eight distinct kinds of gluons. Even more interesting is that photons do not interact with each other but gluons do. The result of this is that, although photons can travel enormous distances in space to produce electromagnetic forces, gluons produce a force that increases in strength the further apart you pull a pair of quarks. This unique feature of gluons is what confines quarks inside protons and collectively confines protons and neutrons inside the nuclei of atoms.

Certain particles can decay into simpler particles and this also requires a nuclear force. The neutron can exist for an average of ten minutes before it decays into a proton, an electron, and a neutrino. Particle decays are a signpost of a third force in nature called the weak nuclear force. Just as the electromagnetic force is carried by the exchange of photons and the strong force is carried by gluons, the weak

Above *Base jumping relies on the force of gravity to ensure that the participant has an exciting trajectory, which is the history of the participant through space. In relativity, this four-dimensional trajectory in spacetime is called a worldline.*

force is carried by three particles called Intermediate Vector Bosons (IVBs). Without the weak force, stars could never fuse hydrogen into helium to support themselves. Also, supernovae would never be able to detonate to enrich space with new elements like carbon, oxygen, and iron.

Gravity, the fourth elementary force of nature, has been known to humans for literally millions of years, but only studied in detail during the last 400. It is, paradoxically, the weakest force in nature but at the same time the most ubiquitous. Every scrap of matter in the universe from quarks and electrons to stars and galaxies produces this force, which only causes attraction. The essential nature of gravity was worked out by Sir Isaac Newton in 1666, who showed how its effect behaved with mathematical precision. Like electromagnetism it is an inverse-square law force that diminishes with separation between material bodies. This also explained the regularities among the planets in the solar system, including the reason why they orbit the sun in elliptical paths. This idea that gravity was just another kind of force was completely changed by Albert Einstein in his investigation of relativity. Einstein's general theory of relativity published in 1915, offered a completely different explanation for the existence of gravity that also included a new way of thinking about space and time.

Einstein's relativity theory says that the correct arena for describing how objects move is called spacetime. This is a four-dimensional plenum that includes the three-dimensions of ordinary space alongside the one-dimension of time. Only by treating an object's evolution as a path through spacetime, called a worldline, rather than just space by itself, can you accurately account for its movement and behavior. Worldlines represent the history of a particle as it moves through space. What Einstein's theory of general relativity also showed in its mathematics was that spacetime can be curved, and this curvature would be experienced as what we call the gravitational force. The shortest path of a planet through the curved spacetime produced by the more massive sun, its worldline, would look like an ellipse in three-dimensional space, and a corkscrew helix in four-dimensions with the axis of the helix along the time axis. In attempting to travel along the straightest possible corkscrew worldline, the planet would experience this curved spacetime as the ordinary force of gravity just as Newton had prescribed. So gravity is not a force in the same way that the strong, weak, and electromagnetic forces are, but is a consequence of particles moving in a curved or warped space. There is also no experimental evidence that gravity is produced by the exchange of force-carrying particles such as "gravitons," although there is a lot of theoretical evidence that they probably are.

Above Spacetime Warped by a Black Hole *is an artistic rendering by Mark Garlick, which also illustrates how worldlines are distorted by the curved spacetime near a physical object.*

// Ingredient 3: Hidden fields and forces

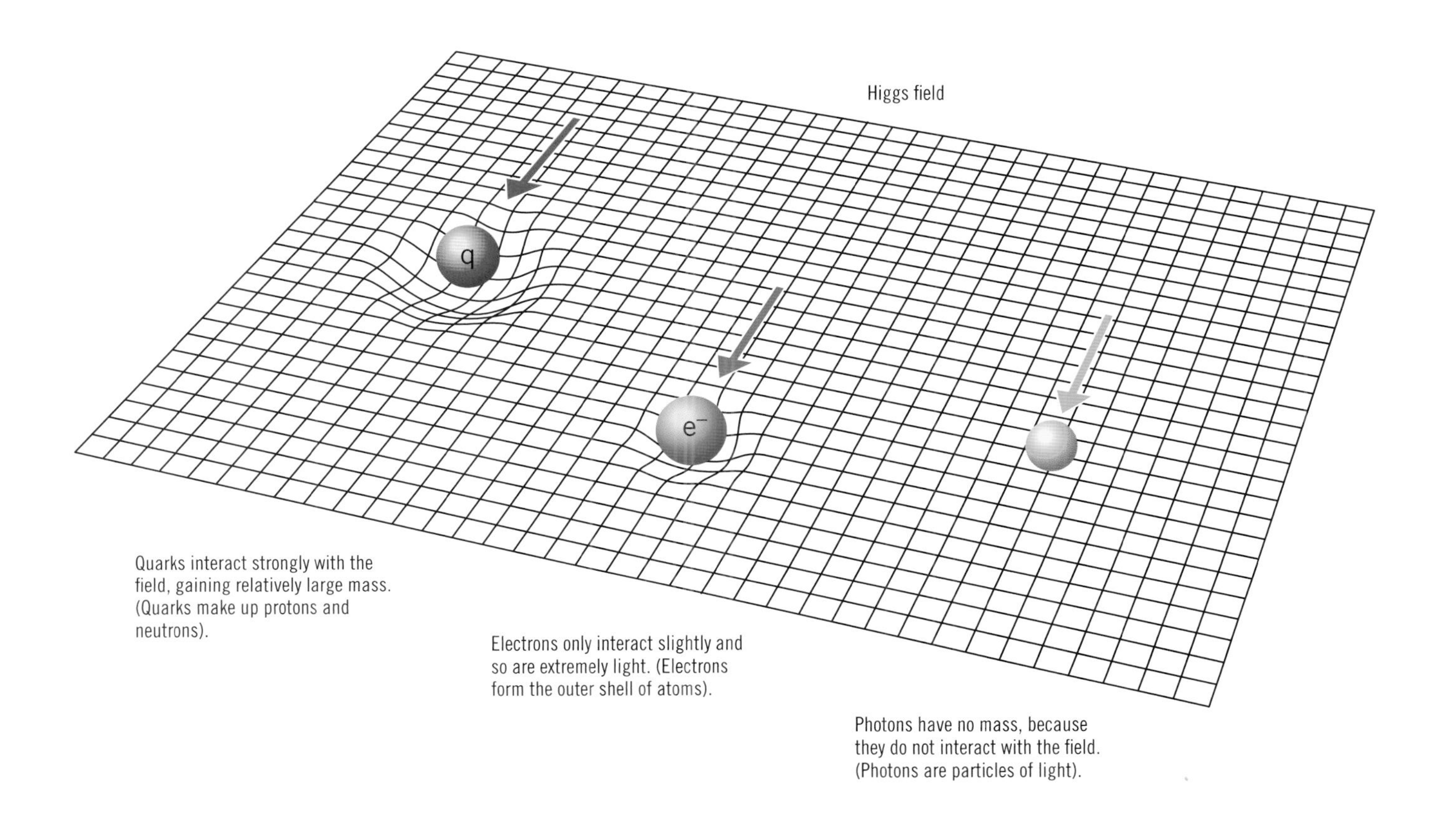

Missing from the previous discussions of forces and matter is a critical 25th particle called the Higgs boson. Unlike the 12 force-carrying particles that produce the strong, weak, and electromagnetic forces, this field would quite literally be hidden in "empty space." Its interaction with the other matter and force-carrying particles in space would give all of the particles in the Standard Model the property of mass we measure. Photons and gluons do not interact at all with the Higgs field and have zero mass. But electrons and neutrinos interact weakly with the Higgs field and gain a small amount of mass. Quarks, muons, and tauons interact still more strongly with the Higgs field, and finally the IVB particles interact the strongest of all. When a force-carrying particle gains mass, the force that it creates does so across a shorter and shorter range until finally for the massive IVB particles their ranges collapse to a size smaller than an atomic nucleus. In 2012 LHC physicists announced the detection of the particle associated with the Higgs field called the Higgs boson at an astonishing mass equivalent to just over 130 protons but carried by a single elementary particle.

Above *One way to think of the Higgs field is that it acts like invisible molasses in space to make it harder for particles to move. This reduction in movement, which is actually an increase in the particle's inertia, is seen as the quality we call mass.*

The existence of the Higgs boson had been predicted nearly 50 years earlier by physicists Peter Higgs and François Englert who also received the 2013 Nobel Prize in physics for their work. But long before its discovery at the LHC, this particle and its associated field had already become a staple of many advanced theories of unifying the forces in nature.

Although the particle associated with the invisible Higgs field embedded in empty space has been discovered, the same cannot be said for the second, even more mysterious, field of nature—"dark energy." Measurements of very distant supernova explosions show that they are fainter than they

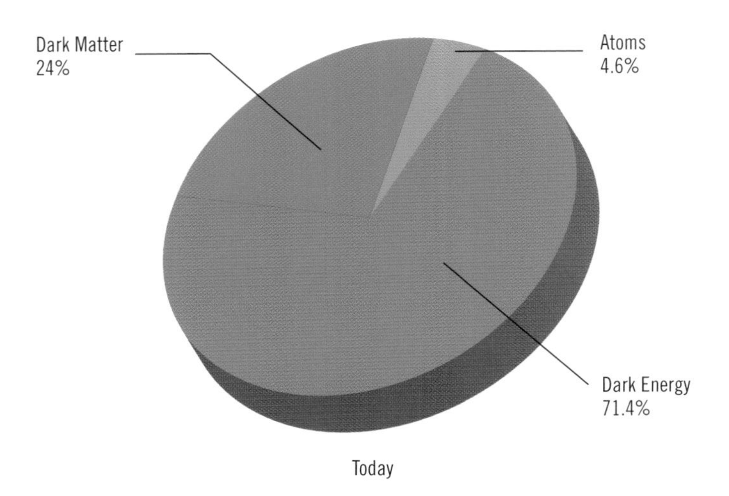

should be if the expansion of the universe was constant. They appear to us as though they are further away, which could only happen if the recent expansion of the universe were accelerating. Another way the effects of dark energy can be detected is by carefully studying the fireball light left over from the birth of our universe in the Big Bang. Spacecraft such as NASA's Cosmic Background Explorer (COBE) and the Wilkinson Microwave Anisotropy Probe (WMAP), along with the European Space Agency (ESA) Planck satellite, used very sensitive radio receivers to measure the brightness of the cosmic microwave background radiation (CMBR).

Left *When the universe is "weighed" using CMBR data from the WMAP and Planck spacecraft, astronomers find three kinds of energy that contribute to the cosmic gravitational field. Surprisingly, ordinary matter in the form of stars and gas is only 4.6 percent of this total.*

Below *Instead of the separations between galaxies increasing linearly in time such as 2, 4, 6, 8... the separations are changing at an exponential rate: 2, 4, 8, 16, 32....*

They discovered that it was very smooth across the sky as predicted by Big Bang cosmology but there were very minute irregularities. When compared with detailed mathematical models of the evolution of these clumps, the amount of dark energy, dark matter, and ordinary Standard Model matter could be deduced. The result of these WMAP and Planck measurements of the CMBR is that the universe consists of 4.6 percent ordinary matter, 24 percent dark matter and 71.4 percent dark energy. This invisible dark energy field permeates space and gets stronger as the volume of the universe grows bigger, which causes the expansion of the universe to accelerate. In the unimaginably far future 100 billion years from now, this invisible dark energy field will actually cause galaxies to dissolve, planets to shatter, and even atoms themselves to fly apart in what cosmologists call the Big Rip.

// Ingredient 4: Space

To build a universe we need the previous three ingredients, but we also need a place to put them. That place is usually called space, which in our universe has exactly three dimensions and is unimaginably vast in scale. For thousands of years, space was considered to be a passive container into which the ingredients of the world were placed. Even as late as the 18th century, Sir Isaac Newton considered space to be a fixed, passive, absolute, frame of reference for all things in the universe so that their locations and movements could be rationally and consistently described via mathematics. But by the early 20th century, the success of Einstein's special and general theories of relativity demolished the need for such a pre-existing and eternal Newtonian space. Instead, Einstein proposed that space was actually a human fiction. It was part of a more complex physical object that also included time as a fourth dimension, creating what is called "spacetime." In fact, spacetime in Einstein's new relativistic theory for gravity was just another name for the gravitational field itself. This is because the same mathematical symbol used to describe the geometry of spacetime is also used to describe the strength of the gravitational field. Since matter and energy create gravitational fields, space was created by matter and energy

Abell 2744, located four billion light-years from Earth is situated in the constellation Sculptor. The giant galaxy cluster is about four million light-years across.

as well. So, space does not pre-exist as a passive container for matter and energy in the universe. In fact, the birth of our universe also brought into existence space as well. Space in the cosmological setting is nothing more than an interval of distance between objects. According to general relativity, it is not a physical "thing" whose rate of increase is limited by the speed of light. Objects such as galaxies are embedded in space, but their motions are slower than light-speed while the space between them can dilate faster than light.

Once you have the first three ingredients, matter, forces and hidden fields embedded in empty space, you get space itself free-of-charge! Amazingly enough, the same may be true for that mysterious thing we call "time."

Above *General relativity says that space can be warped in the presence of matter, shown in this artist's illustration of the geometry of space near a star.*

// Ingredient 5: Time

The most daunting mystery of our existence is the nature of time. For thousands of years, philosophers of all stripes and persuasions have tried to explain what this "thing" is that seems to inexorably flow along from past to future, and organizes our lives as well as events in the entire physical universe. This has led to a small number of key insights that seem on the face of it very intuitive. The most widely known of these is what Sir Isaac Newton considered a cosmic "master clock" that counted the instants of time, which flowed past-to-futureward independently of the behavior of matter. Einstein's relativity theories, along with countless experimental results, proves that there is no such synchronized Cosmic Time. Every nugget of matter in motion has its own master clock called proper time, and proper times cannot be synchronized across the universe to create a Cosmic Time. Time, like space, is only defined by relationships between collections of matter (clocks and other objects including even fields themselves). This forces us to look more closely at the nature of spacetime itself.

Spacetime is the bedrock of our entire universe, but it is defined by the innumerable worldlines that begin and end within its four-dimensional volume like the spaghetti at the bottom of a strainer. Each worldline is a collection of events that are connected by some cause-and-effect relationship forming the history of a particle. Events are specific instants in time and locations in space where one particle has interacted with another, such as when a photon of light is emitted or absorbed by an atom, or when you meet someone at a specific date and time at the base of the Eiffel Tower in Paris. The events along each worldline are specific instants in time carried by the clock associated with that worldline, much as you measure your current time with a wristwatch that you carry with you. This local time is called proper time according torelativity. According to Einstein's theory of general relativity, it is the shapes of the worldlines that collectively define the geometry of spacetime, not the other way around. From the vantage point of seeing time and space together as spacetime, the entire history of a particle represented by the innumerable events along its worldline is seen all at once. Spacetime does not evolve in time, it simply exists in its totality. This implication of relativity is called the Einstein Block Universe, and presents physicists with a significant problem: if the history of an observer is represented by their worldline from the instant they were born to the instant they died, how is the current moment in time you are experiencing, called Now, singled out? This is also a reflection of the fact that all equations in physics that model the motion and evolution of an object or system do so by invoking a variable, t, that represents time. But nowhere in any of these mathematical models is t=Now singled out as a special moment.

Even more confusing is that we do not experience time the way we experience the three dimensions of space. In fact, the only way we experience time and keep track of our Nows is by consulting other nearby collections of worldlines called "clocks" that keep proper time and help us keep track of changes in our environment. If nothing in our environment changes, time does not exist in a meaningful way at all. A more remarkable feature of the spacetime view is that time is not a feature external to the ensemble of worldlines that make up the contents of our universe. Time emerges and exists only within the spacetime of our universe.

Although we have gone far beyond ancient philosophers in understanding how time and space are intertwined, we have not managed to solve the Problem of Now, as physicists call it, or why time exists at all. What we do know is that, just like space, time came into existence at the Big Bang and according to Einstein's theories of relativity is a feature of matter and energy. So time (and space) does not need to be created as an independent ingredient before matter appears, but comes into existence alongside space and matter.

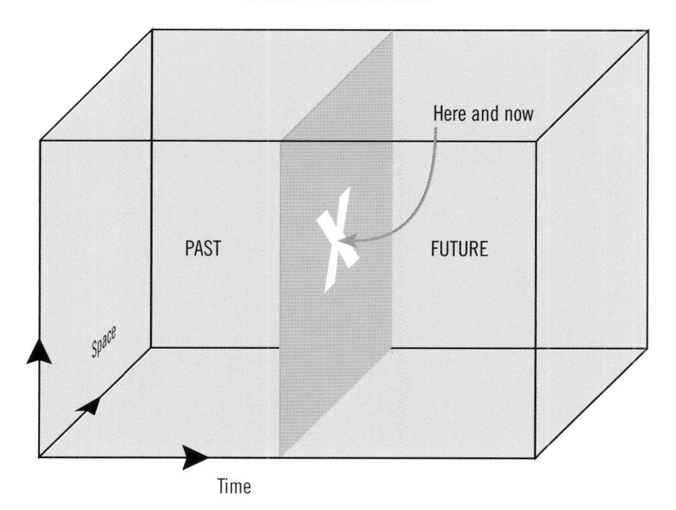

Left *A small portion of the block universe of relativity. What we call Now is a slice through this block at a particular instant in time. The slice represents a snapshot in time of where all objects are in three-dimensional space at that exact moment in time. Physicists cannot explain why Now is more important than any other moment in cosmic history.*

Time is one of the most puzzling features of our world, but with Einstein's relativity theory we can now think about it as a co-equal partner to 3D space within the arena of spacetime. Our differing perceptions of space and time are not relevant.

// Ingredient 6: Laws of nature

Once we have the ingredients to the universe and the space and time to contain them, we are left with a thorny issue: how, exactly, do we describe how these ingredients will interact? For instance, gravity obeys the inverse-square law, which says that as you double the distance between objects, the gravitational forces fall to a quarter of their previous strength. If you triple the distance, the force becomes 1⁄9th of its former strength, etc. But it is one thing to say that gravity obeys the inverse-square law, but why this law and not some other? Along with stating the specific laws, we also have to specify specific universal constants such as the speed of light, or the mass of an electron among many others. For example, one simple law of nature involves the gravitational force and is represented by the following equation developed by Newton, which also introduces a new fundamental constant in nature called G.

$$F = -G\frac{Mm}{d^2}$$

Laws of nature include a bewildering collection of precise mathematical statements that prescribe exactly how forces should behave across space and time, how electrons behave inside atoms and how gas pressure, volume, and temperature are related. During the last 200 years, entire fields of physical science have emerged, each with their own collections of laws. Sometimes these laws cross over from one field to the next. For example, in quantum physics, a set of laws has been discovered for how electrons behave. Because electrons are critical atomic components for describing a variety of laws in chemistry, chemistry and quantum physics are intimately connected. Quantum laws help chemists predict the behavior of new molecules and also explain many laws of chemistry deduced from experiment going all the way back to the alchemists of long ago.

Laws of nature often have to be based on a set of physical constants that have the same values across many different phenomena. For example, Newton's universal constant of gravity, G, has a numerical value that is the same if you are predicting the motion of a baseball in flight, or the movement of vast collections of galaxies in deep space. The speed of light, c, has the same value everywhere and by all observers, whether they are waiting for radio waves to arrive at their cell phones or are traveling at very high speeds.

A few universal constants in nature

Quantity	Symbol	Value
Gravitational Constant	G	6.67430×10^{-11}
Planck's constant	h	$6.6260755 \times 10^{-34}$
Speed of light	c	2.99792458×10^{8}
Electron charge	e	$1.602176634 \times 10^{-19}$
Stefan-Boltzmann constant	σ	5.670374×10^{-8}
Permittivity of free space	ϵ_o	$8.854187817 \times 10^{-12}$
Boltzmann's constant	k	1.380649×10^{-23}

For our universe, the collection of laws and universal constants we have arrived at through experiment and observation appear to be the same regardless of your location in the universe, but we seem to exist on an incredible knife's edge. If these laws and constants were different, it is usually the case that our universe would look very different than it does. It is even likely that life itself would be made impossible by the changes. For example, nuclear forces are essential for allowing atoms to exist, and for nuclear fusion to power the stars. In our universe, the strong nuclear force is about 100 times stronger than the electromagnetic force, but if it were slightly stronger, say 150 times stronger, fusion reactions in stellar cores would liberate more energy and stars would explode. If the electromagnetic force were stronger than it is, chemistry would require more energy to form molecules or break them up. A slightly weaker force would make complex molecules beyond water impossible to form before they literally shook themselves apart. We live in a very hospitable universe where these two forces are very conducive to having stars that live a long time, and chemistry involving carbon atoms leads to the complexity of organic life and DNA.

We don't actually know where the physical laws came from, but one possibility is that they emerged soon after the Big Bang as the universe evolved. None of the known laws make any sense unless you already have in existence space and time. In fact, these laws may not even be fixed in time but, like matter and space, may actually evolve over time. So in some sense, even the laws of nature are not ingredients we need to specify in advance but come into existence once matter, space, and time appear. Nevertheless, the origin of the physical laws and constants remain a mystery to physicists today.

Opposite *Science has discovered many mathematically-precise laws that underlie nearly all of the phenomena we observe in our physical universe.*

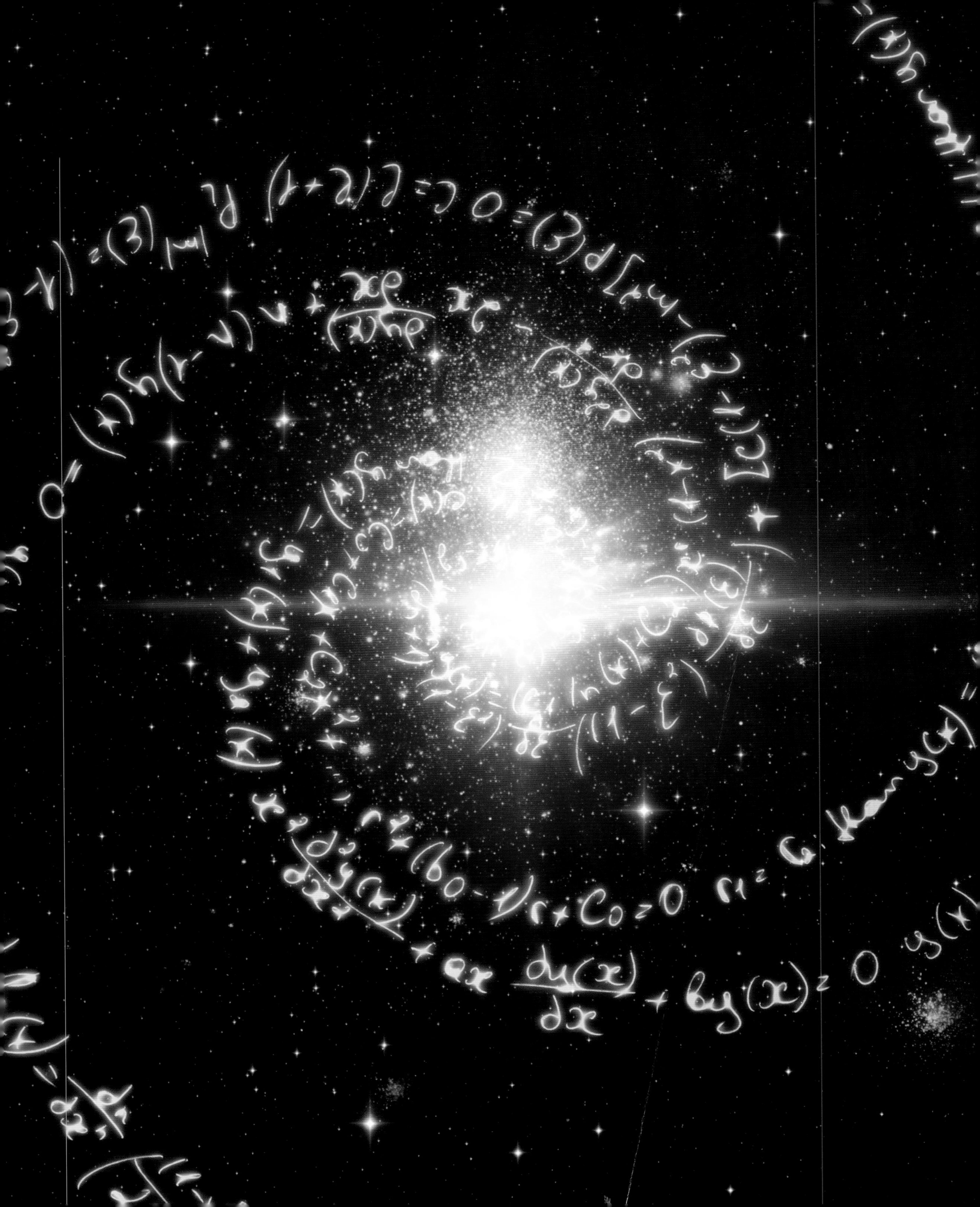

// Ingredient 7: The multiverse

An artist's rendering of a portion of the multiverse; a hypothetical arena in which all possible universes exist.

With all of the possible laws of nature and values for the physical constants, why does our universe have the particular ones that it does? Could there be some super law of nature that only allows ours to be the ones selected because of some deep internal logical consistency not shared by other universes? Our universe could be absolutely unique with no other kinds of universes possible that lead to logically consistent phenomena spanning billions of years and culminating with sentient life. A second possibility, called the multiverse, is that all of these other possible universes actually exist. We find ourselves in this one because it was the one capable of nurturing life and consciousness. The laws and constants seem strange only because we are here to experience them and marvel at their

uniqueness. Other possibilities for these laws and natural constants would lead to lifeless or still-born universes. In either case, we can legitimately ask the even more confounding question: "Where did the multiverse or the super-law of nature come from?"

There may be universes "out there" that have quarks but no electrons. There may be some that look otherwise identical to ours, but where the only difference is that the speed of light is 177,323 miles per second (285,374 kilometers per second) and not 186,282 miles per second (299,792 kilometers per second) as it is in our universe. We cannot interact with them in any way because they are so far away from our universe, or perhaps even inaccessible across a gulf of spatial dimensions through which we cannot travel or glean information. This is a very uncomfortable situation for scientists because there would be no experimental or observational way to detect them and so their existence is literally beyond the Scientific Method to prove or disprove. All we do know is that our very existence may somehow be part of the explanation for why our universe looks the way it does; why the matter and forces behave as they do; and why the forms and values for the laws and constants of nature are what they are. In a very real sense, the existence of life and humans is a hidden eighth ingredient to our list.

Now let's put all of these ingredients together and see how they lead to the universe we see around us!

Below *Light's finite speed determines how far we can see, but as the visible universe (blue sphere) grows, it is an ever-smaller part of a larger universe expanding even faster.*

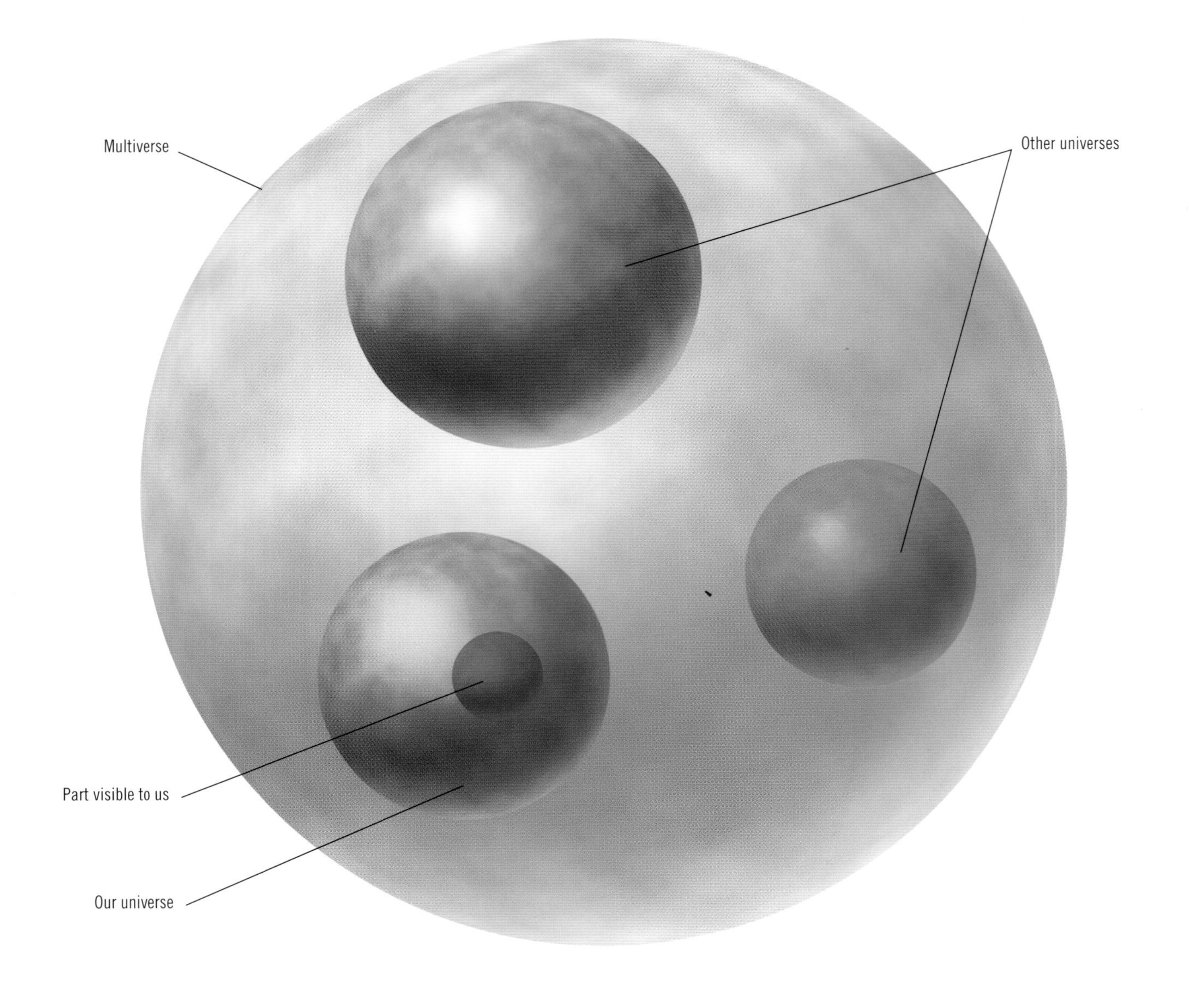

EXPLORING THE BIG BANG

The previous list of seven ingredients makes for a complex cast of characters from which to fashion the modern story of our universe. Scientists have worked for over a century to mix these ingredients together into a deep understanding of how our universe formed and has evolved since its inception 13.8 billion years ago. All creation stories in human experience have a beginning in time. What happened during this moment is surrounded by a bewildering series of dramatic changes, and a redefinition of what we mean by the actual event itself. As it turns out, it is not a single event in time and space, but a myriad of equally important moments. All of these details are now the province of a scientific subject called cosmology.

Although it is impossible to accurately render what the Big Bang looked like, artists remain undaunted in trying their hand at showing our universe emerging from an event that created time, space, and matter.

// Time zero

When do we know we have arrived at the start of our story of the beginning of our universe? For many people, this story begins with their own birth and nothing else has any practical consequence. For most of human history the time horizon for the essential history of our world has been a few human generations or at most a few thousand years. Once the scientific approach to studying the universe commenced 400 years ago this horizon was steadily pushed back in time by millions of years with the discovery of dinosaur fossils, and then billions of years with studies of the evolution of our sun. For the universe-at-large, cosmology provided a framework for exploring

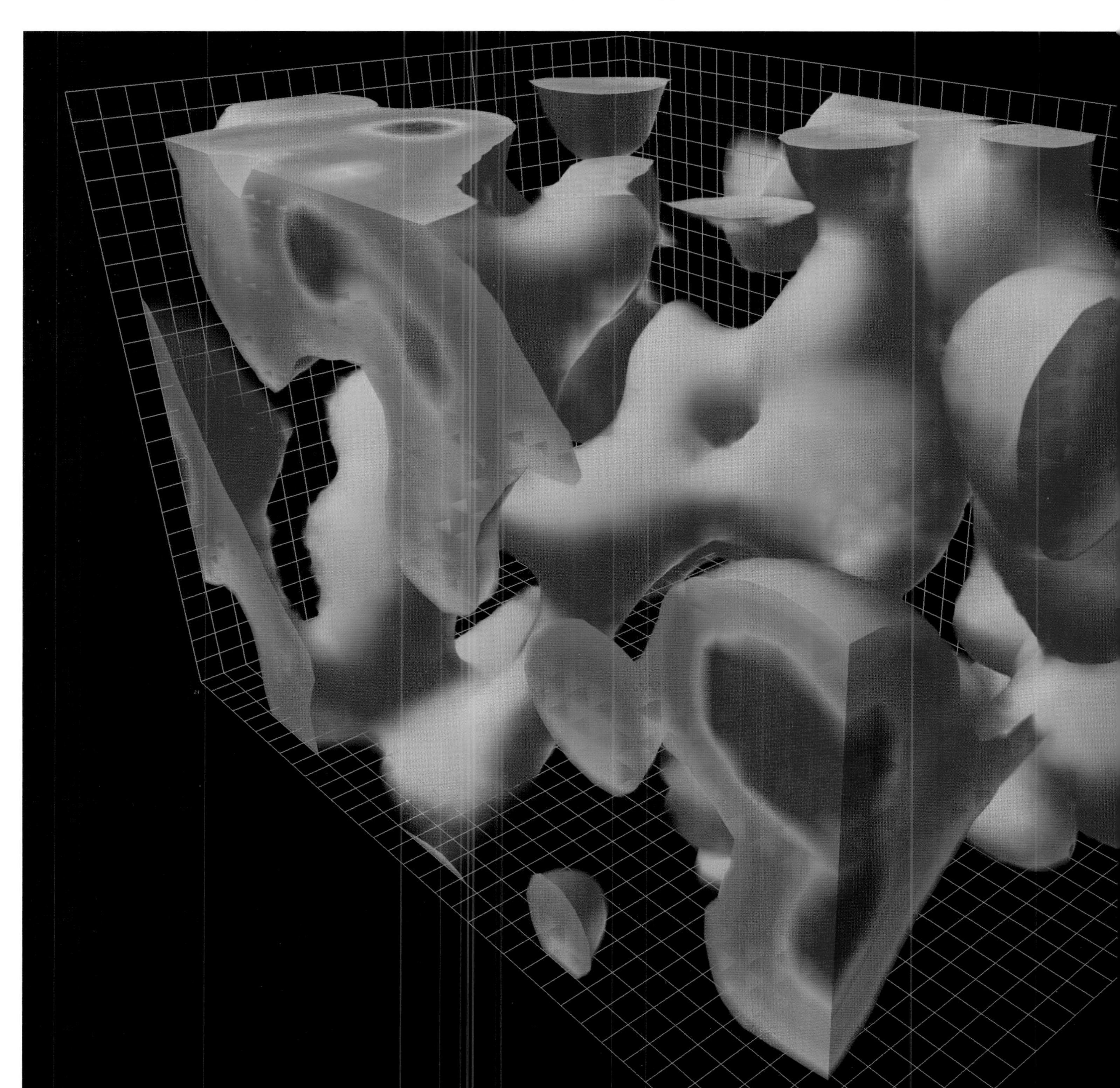

the behaviour of matter as the universe aged. Physicists meanwhile explored nuclear and quantum physics, laying out principles and laws that showed how matter behaved at ever-increasing energy. At first, during the 1950s, it sufficed to probe the early history of the universe when it was a seething billion-degree plasma that produced the primordial elements hydrogen and helium at a time about 20 minutes after its birth. But in the decades to follow, this horizon in time was breached as the Standard Model was discovered, codified, and explored at facilities such as the Large Hadron Collider, among others. Now armed with experimental data, we can push this horizon back beyond the first microsecond to an astonishing trillionth of a second after the origin moment, and still we have not exhausted the story of how our universe came to be. The answer to the children's question "Are we there yet?" remains an astonishing and emphatic "No."

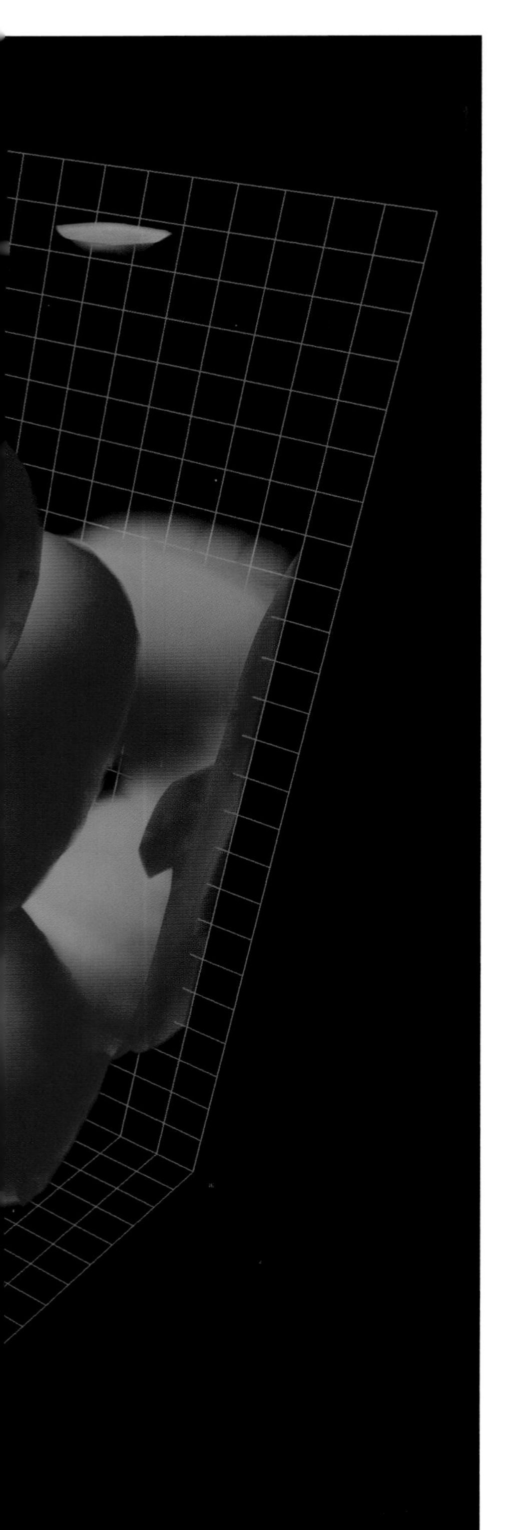

The origin of the technical story of how we think our universe came to be is based on a mathematical framework developed by Albert Einstein for describing gravity, space, time, and matter under extreme conditions. Soon after its publication in 1915, Einstein's theory of general relativity was applied by astronomers to describing the universe, only to discover that we live in what must be an expanding universe in space and time. This expanding universe also had a definite origin in time, called by the English cosmologist Sir Fred Hoyle the "Big Bang;" an unfortunate moniker that caught on and persists to the present time. Big Bang cosmology, more properly termed Friedman-Robertson-Walker cosmology, has been developed and refined by hundreds of physicists and astronomers so that it now provides us with a detailed mathematical framework for describing how the temperature, density, and scale of the universe changed after Time Zero. In a very literal sense, we take the expanding universe "movie" as we see it today when the universe is 13.8 billion years old spanning billions of light-years, and mathematically run the movie backwards in time. The mathematics allows us to calculate its scale, temperature, and density as far forward in time as we feel like taking this journey. We can even run the Big Bang in reverse all the way to Time Zero (t=0) itself, but there is an important caveat: because Big Bang cosmology also describes both the origin of space and time, it does not allow us to ask what happened "before" the Big Bang itself. But after this instant, we make immediate contact with all of the ingredients covered in the previous chapter, from the entire gamut of the known Standard Model particles and forces to the enigmatic contributions by dark matter and dark energy. But first we have to confront the millennia-old conundrum of where everything came from in the first place.

SOMETHING FROM NOTHING

It has been a deep challenge for humans to imagine how something (our universe) could have appeared out of nothing since this problem was first debated in the ancient text of the *Rig Veda* more than 3,500 years ago. A surprising number of these stories seem to begin with the artifice of "dark formless waters." It may come as a surprise, but in the modern age of cosmology this problem is not nearly as troublesome. What is now at issue is, which one of a variety of something-out-of-nothing scenarios is the correct or more accurate one? Since the 1920s, there are now at least two mechanisms known from investigations of atomic physics that may offer a clue, and many more theoretical ideas that currently have no experimental evidence by which to evaluate them other than the very existence of the universe itself. In one humorous idea, our universe was the result of a lab experiment by a graduate student gone wrong in another universe.

Left *What we call empty space is actually filled with a patina of quantum fields including gravity, whose fluctuations can bring into existence matter and energy seemingly from out of nothingness.*

// Vacuum fluctuations and quantum tunneling

Studies of the structure of the atom and atomic physics have revealed a world governed by quantum laws that hinge on how an observer or measuring device affects the states of electrons and other particles. One powerful principle developed by the German physicist Werner Heisenberg is called Heisenberg's uncertainty principle (HUP). Here's how it works.

The German physicist Max Planck came up with the idea that light is carried by particles we call photons and Einstein experimentally confirmed this idea in his study of the photoelectric effect. Einstein's special relativity also says that energy and mass are equivalent physical quantities via his iconic equation $E=mc^2$. Now we run into an interesting problem when we use HUP.

When you try to observe where an electron is located in space, you use a photon that collides with the electron and you measure the electron's location by measuring the reflected photon. To make a precise position measurement you use a photon of the shortest possible wavelength. But this photon carries a lot of energy. When it reflects

Below *An artist's rendering of vacuum fluctuations in empty space.*

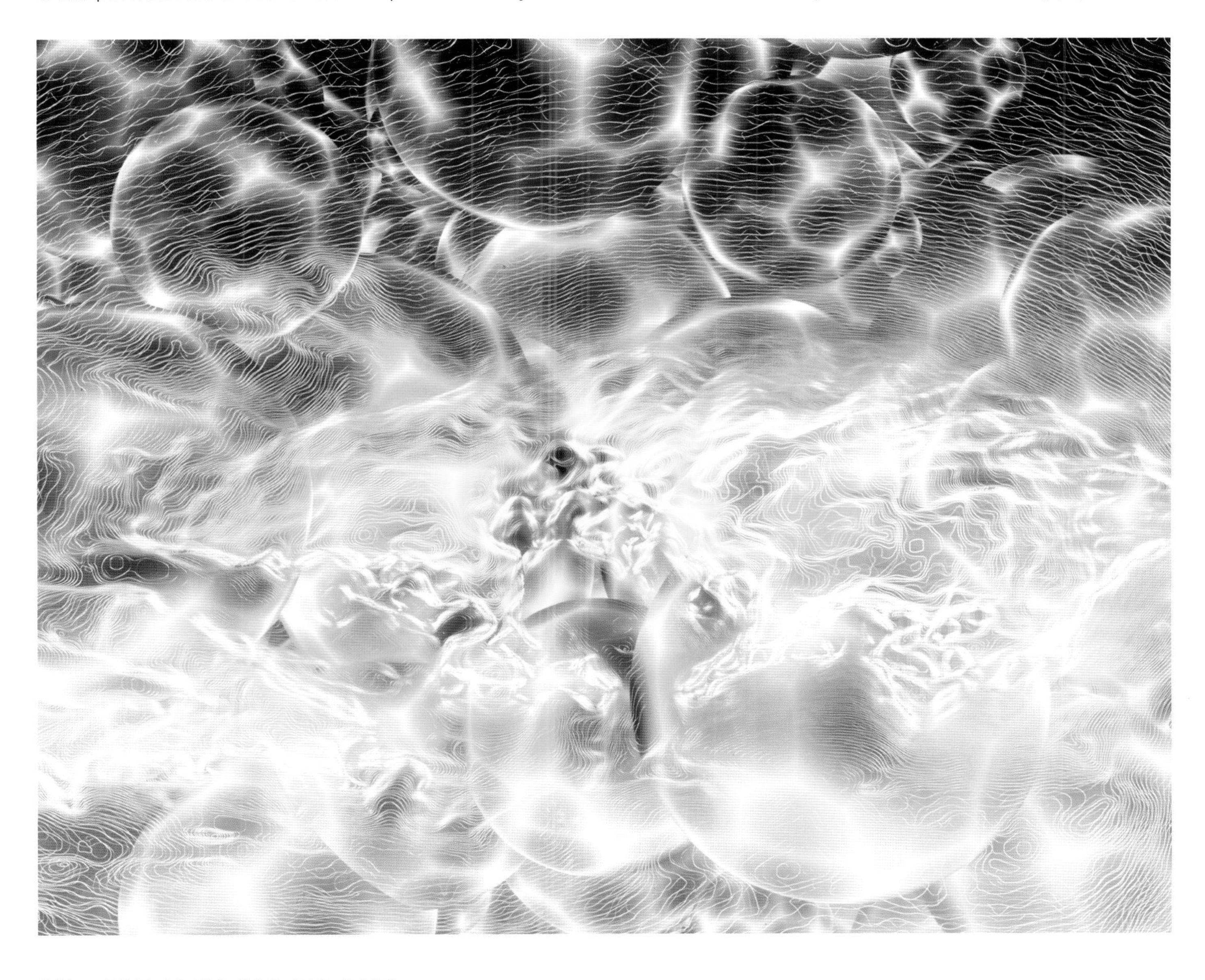

QUANTUM TUNNELING

In the quantum world, even matter takes on wave-like properties. Electrons can be thought of as either point particles, which is easy for us to visualize, or as wave-like objects that are spread out in space. A number of important experiments in the 20th century have confirmed this bizarre quality of matter and it also leads to another important phenomenon that involves the energy of these particles called quantum tunnelling. To see how this works, we can think of a rocket trying to leave Earth. If the rocket does not have enough speed (energy), it cannot leave Earth's surface because of gravity. It will travel upwards a little way, but then come crashing back to the ground. In Newton's physics, an object has to surmount an energy barrier (defined, for example, by a planet's escape velocity) in order to become a free particle and, in the rocket example, leave Earth. In the atomic world, where particles have wave-like attributes, the situation is more complicated.

Because of HUP, when a particle gets close to the energy of a particular barrier at a specific point in space, it can actually take advantage of its wave-like properties and sneak through that energy barrier. At its specific location, we can't tell with high precision whether its energy is exactly below the barrier energy or exactly above the barrier energy. Because of this confusion, the particle can "tunnel" through the barrier and get outside even though this is not permitted by Newton's physics. This effect is what allows the sun to shine. It is also the cornerstone principle for many modern devices such as cell phones. For cosmology, it means that empty space at one energy (E=0) can become filled space at another energy (E > 0) by quantum tunnelling from one state to the other. The time it takes is very long when the energy difference is large, but can be incredibly short if the energy difference is small. This is why some radioactive isotopes have very long half-lives while others have very short half-lives for decay. Physicists think of the seemingly empty space of our universe as a system that could have its own energy, but that this energy might not be the lowest possible energy it could have. This means that empty space could tunnel into a lower-energy state, with devastating consequences for the stars, planets, and life in our universe.

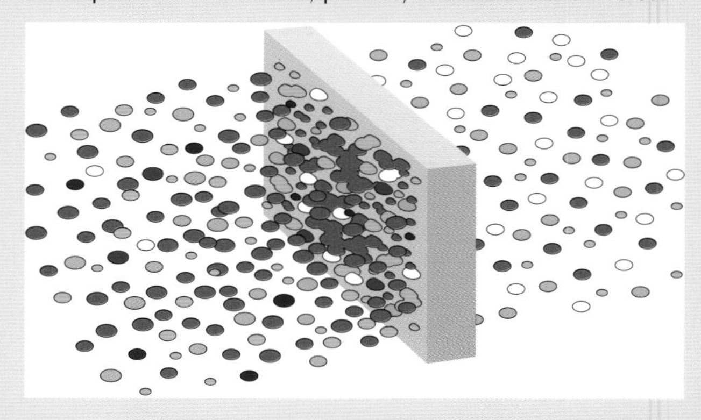

Above *In every chemical reaction, what stands between the reactants and the products is an energy barrier. Quantum tunneling can lower the amount of energy required for a reaction to proceed.*

off the electron, the collision causes you to lose precise information about the speed of the electron. So, the better you know the position of the electron, the worse you know its speed. A similar relationship exists for energy and time. The better you know the energy of a particle, the worse information you have about how long it had that specific energy. In other words, you cannot make your uncertainty in measuring the energy and time measurements both vanish at the same time.

Because special relativity shows that energy and mass are related by the formula $E=mc^2$, this means that you cannot say that a container has exactly a certain amount of mass, or a certain number of particles, unless you observe it for a long, long time. What this means is that particles could suddenly appear and disappear while you are measuring the energy in an empty container. You would not know about it if your measurement took too short a time, and you would swear that the container is actually empty.

We think empty space, a pure vacuum, can be created by just taking all of the atoms and other particles out of a region of space, but HUP says that you can only reach a certain level of emptiness over a particular duration of time. You can never say that the region of space is perfectly empty unless you repeatedly measure its contents for a very long time. In the atomic world this allows entirely new phenomena to occur. For example, you can have an electron and its antimatter partner the positron suddenly appear, and then disappear, quite literally out of nothing, which physicists call a "vacuum fluctuation." This phenomenon has been measured in atomic systems and it is not some theoretical fiction. What this means for Big Bang cosmology is that we have a way of creating matter and energy out of nothing. In actuality, the "nothing" being considered is the existing gravitational field of the universe itself, so fluctuations in the empty spacetime of the cosmic gravitational field in the form of momentary changes in its curvature can spontaneously create matter and energy where none existed before. Truly the creation of something out of nothing.

Cosmogenesis

Some of the earliest ideas in modern physics about the birth of our universe, called cosmogenesis, have to do with the idea of bubble universes. Our entire universe is "simply" one of those HUP vacuum fluctuations that has persisted a long, long time. Suppose you use Einstein's $E=mc^2$ and you annihilated every single gram of matter in the universe into pure energy called its rest-mass energy. The entire rest-mass energy of the visible contents of our universe is a huge number (about 10^{70} Joules), and from HUP leads to an unimaginably short span of time (about 10^{-104} seconds), but there is a missing ingredient to this calculation. In Einstein's general relativity, the gravitational field of the universe also has its own, but negative, energy. If our universe were exactly balanced, the rest-mass energy of stars and galaxies would be canceled by the gravitational energy of the universe giving a net energy close to zero. If that is the case, our universe could be a vacuum fluctuation of almost zero net energy but with a lifetime from HUP of trillions of years.

The question of where the vacuum fluctuation came from is the analogue of asking where vacuum fluctuations inside atoms come from. In the latter case they come from the electromagnetic field embedded in the existing spacetime vacuum of our universe, but for our universe, its vacuum fluctuation would have to come from an even larger spacetime than our own universe. Physicists call this a mother universe, which means that we live in a child universe spawned by the mother universe. But if this is a correct picture, then this same mother universe could have also spawned an infinite number of other child universes by the same mechanism. But the particular vacuum fluctuation in the mother universe would not look like anything you would find inside an atom of our own universe. Instead, the birthplaces for child universes would be inside black holes in the mother universe. This means our own universe, with its own plethora of black holes, could also be the mother universe for its own child universes. According to one idea, every time a black hole forms in our universe it will spawn a child universe, but we can never see this event. The black hole is surrounded by an event horizon that prevents us from observing the birth of these other universes. Although this is an intriguing story based on some ideas in physics related to vacuum fluctuations and black holes, there is almost nothing about it you can scientifically put to a test, so this story is beyond science to prove correct or to disprove.

Another idea for cosmogenesis comes from an attempt to mathematically unify all of the forces into one theory that goes beyond the Standard Model, called string theory. String theory says that the particles in the Standard

Model are not point-like at all but are actually strings of one-dimensional energy that travel through spacetime. We observe them as localized particles because the lengths of these strings are only about 10^{-35} meters—a distance also called the Planck scale. The string versions of the Standard Model particles are confined to our three-dimensional space, which is called a membrane or "brane." This is because mathematically they are

Below *We can imagine our universe existing in a frothy sea of other bubble universes, each created through the action of Heisenberg's uncertainty principle at the quantum scale.*

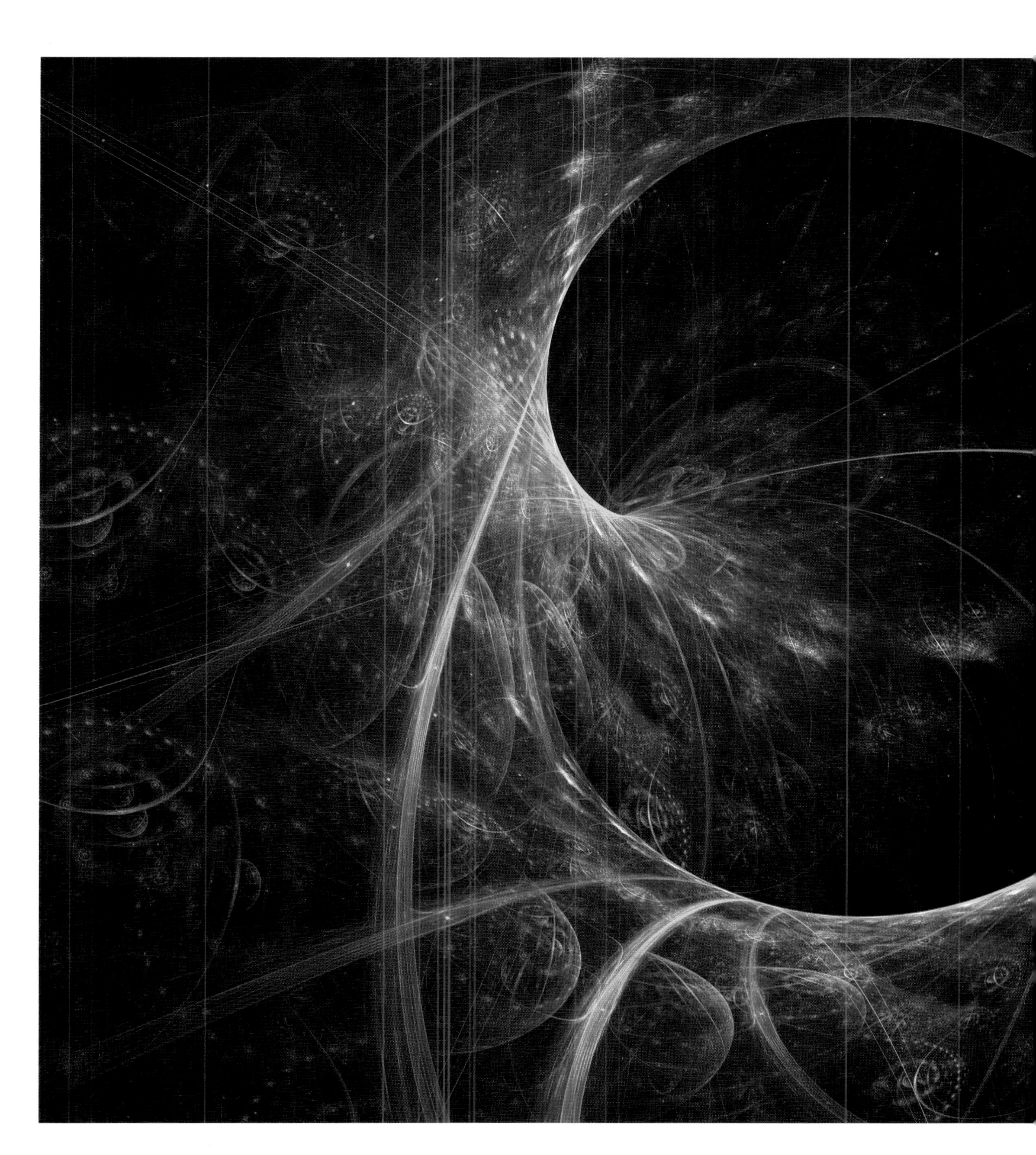

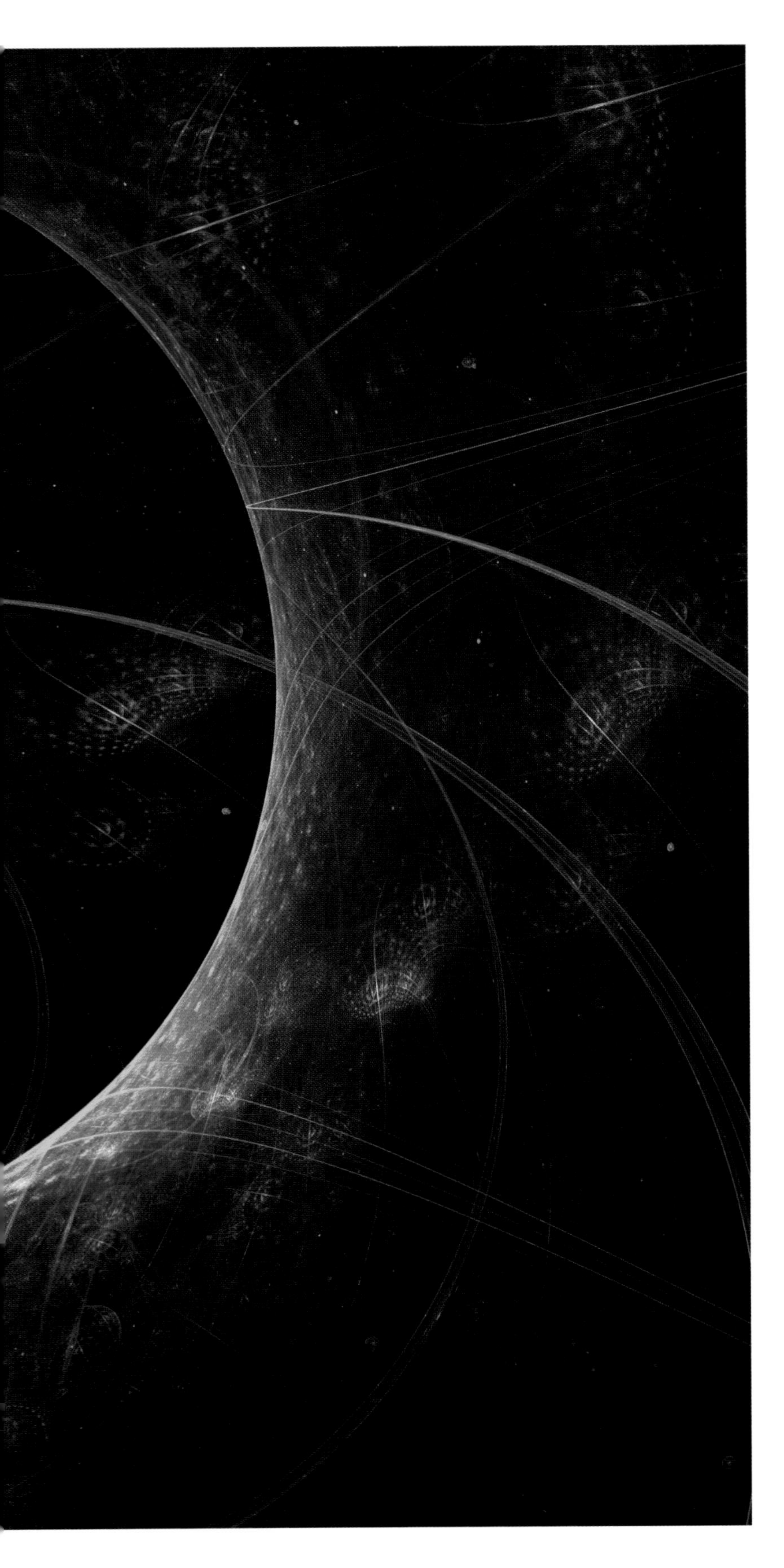

described as loops that have two endpoints anchored in our particular brane universe. But gravity is different. It is mathematically represented by a closed loop and so it is free to leave our brane universe and travel elsewhere. To make string theory work, as well as its extension as a larger theory called M-theory, spacetime has to be 11-dimensional. This 11-dimensional volume is called the Bulk and may contain many other separate brane universes, with gravity free to travel throughout the Bulk. You can think of these other brane worlds as the individual pages in a thick book. They are separated along one of the other dimensions of the Bulk so that they do not intersect, even if each one is infinite in its own three-dimensional space. So how does our universe come into the picture?

One possibility for the Big Bang is that two of these brane worlds were gravitationally attracted to each other and collided. The collision released a huge amount of energy at their point of contact, which set in motion the cosmic fireball in our particular brane world. We call this event the Big Bang. Because in our brane world we were created out of the particles produced by this conflagration some 13.8 billion years ago, we cannot see beyond this distance to the further reaches of our particular brane world. Unlike the previous mother-child universe idea, this idea can actually be put to a test. It all hinges on whether string theory is a correct theory for matter and gravity. String theory relies on a number of phenomena and new kinds of particles that are being searched for at, for example, the Large Hadron Collider at CERN. If the predicted phenomena are not found, string theory cannot be trusted to give us an accurate account of our familiar world, let alone an idea for how to create universes. Thus far, after a decade of study, the LHC has been unable to show that supersymmetry, a key feature of string theory, exists at the energies where it is expected to be an important new feature of how Standard Model particles interact.

The bottom line is that we currently have no idea what started the Big Bang, let alone any way to actually prove that the event happened in the way that we theoretically imagine. Luckily, although we can't say much about what caused the Big Bang, we can say a lot about what happened immediately afterwards. In ancient times, "immediately" meant a billion years but today we can experimentally probe events at 10^{-15} seconds after the Big Bang (let's call this ABB) with some degree of accuracy.

Left *An artist's conception of string theory.*

// The Big Bang

Astronomers have deduced from a variety of observations that the Big Bang occurred about 13.8 billion years ago. General relativity predicts that the origin of time and 3D space for our universe began at the Big Bang. Before the Big Bang there was nothing—not even time itself. Also, we know from our discussion of the contents of our universe that all of the laws of nature and constants of nature were fixed into more or less their present forms soon after the Big Bang. No one knows how this occurred either. If you believe in the multiverse theory, our laws were randomly selected and we are just one of the lucky universes that evolved living, sentient beings capable of asking these questions. There may be uncountable trillions of other universes that got the unlucky combinations of physical constants and natural laws, and are either sterile, didn't exist long enough for stars, galaxies, planets, and organic molecules to evolve, or worse.

With the help of general relativity, Big Bang cosmology starts out with matter and energy spread out in space and rapidly cooling as the universe relentlessly expands. Soon after the Big Bang, when the universe was barely 10^{-43} seconds old, space was filled with a hot plasma of particles and photons being created out of vacuum fluctuations in the "nothingness" of the cosmic gravitational field. Some models of this time predict that these particles were incredibly massive, each perhaps 10^{19} times the mass of a single proton. In fact, they would probably have had the properties of what are called quantum black holes. No technology of today can reach these energies so we have no idea other than theoretical guesses whether such particles existed or not.

This portion of cosmic history is called the Planck era, because all of its essential scales of distance, temperature, time, and density are defined by simply combining three fundamental constants of nature in the right ways: the speed of light, c, Planck's constant, h, and the constant of gravity, G. The temperature of the universe at this time is almost beyond imagining, but it can be calculated to have been about 10^{32}°C. The density of matter and energy could have been as high as 10^{96} kg/m^3. How far could light travel? The furthest distance you could "see" is the age of the universe times the speed of light, which was about 10^{-35} meters. At these densities there would only be one of these supermassive particles taking up the entire three-dimensional space you could see. In

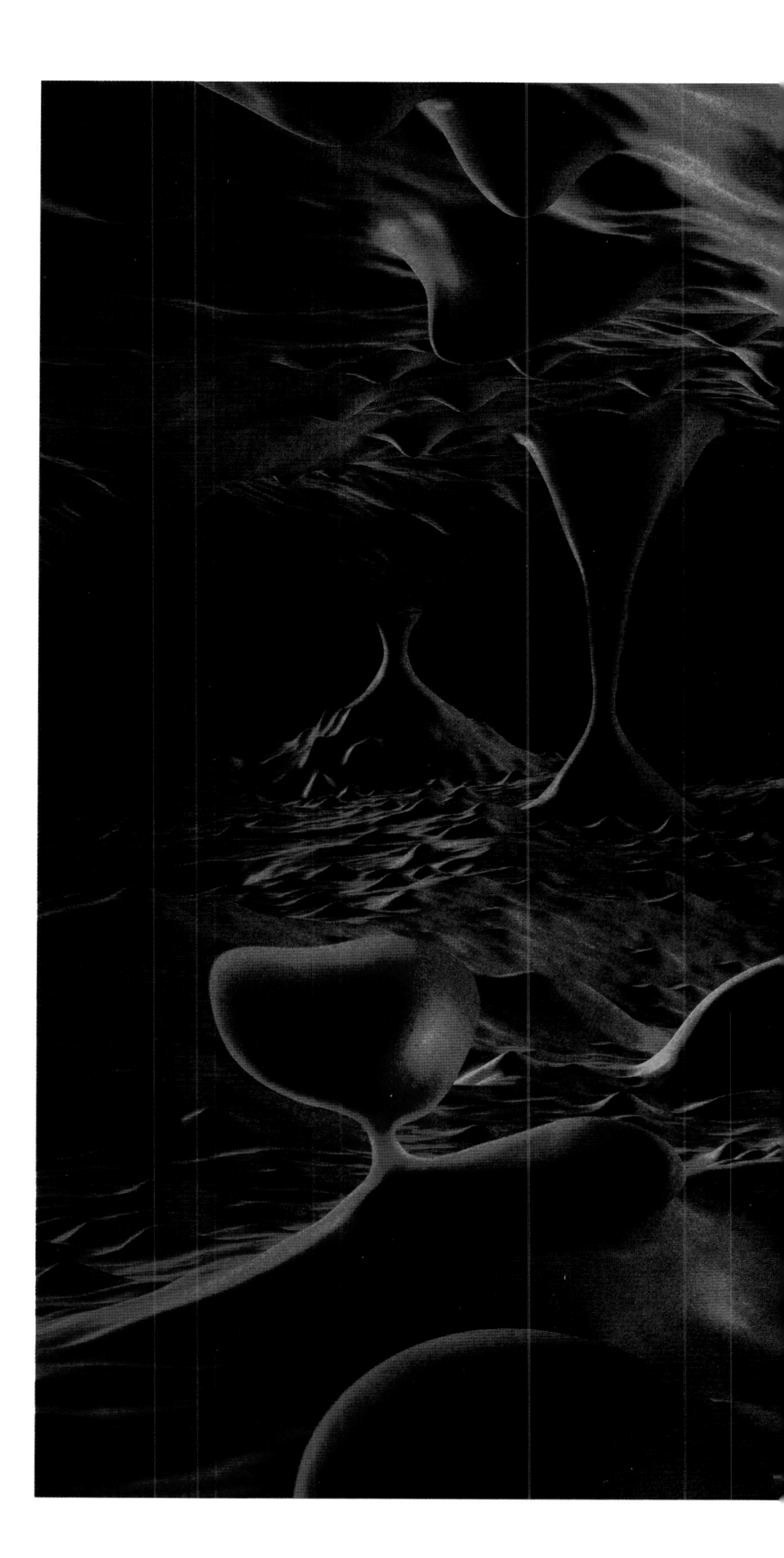

Right *An artist's representation of the primordial quantum foam from which our universe emerged as a bubble from one of these patches.*

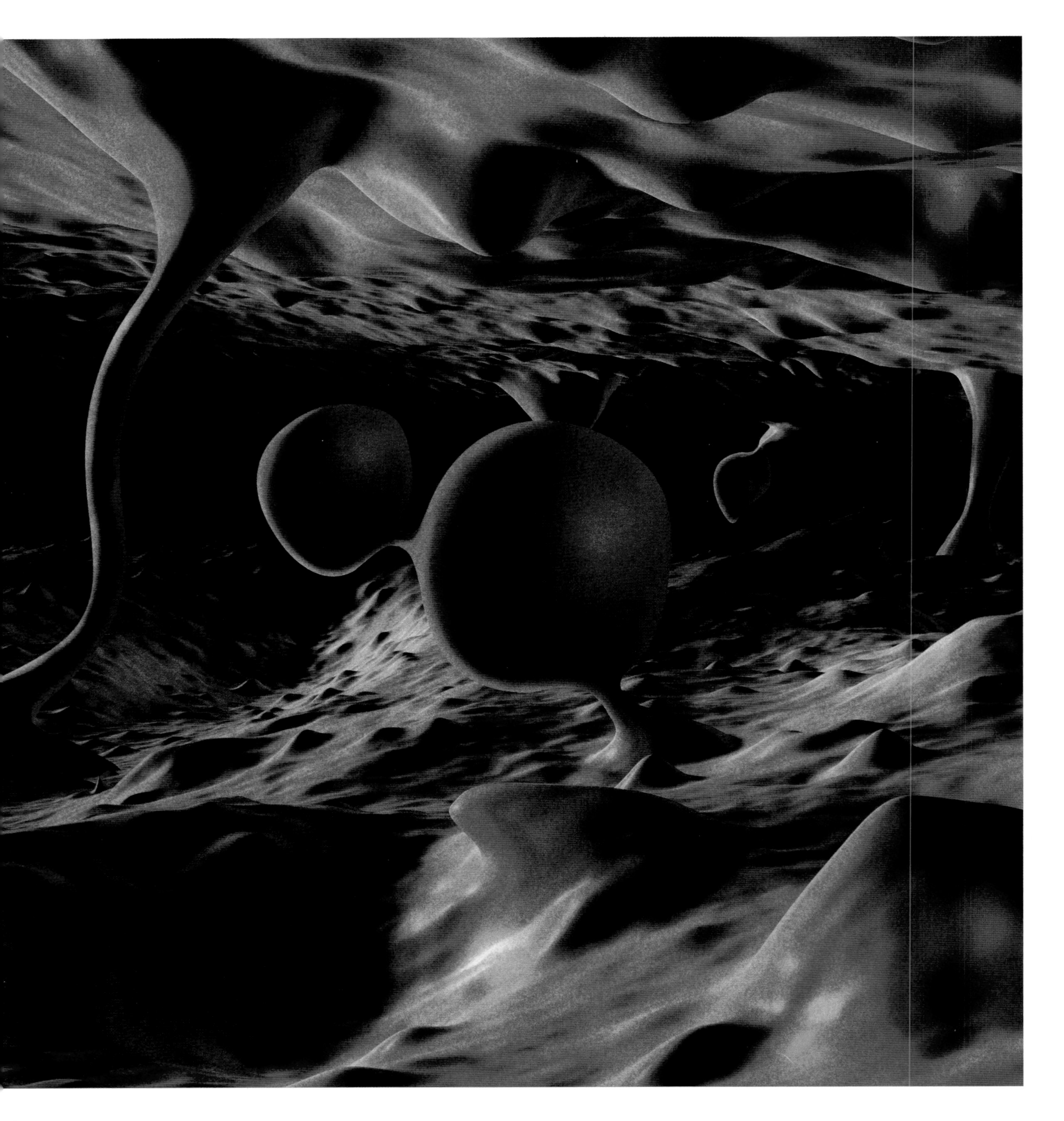

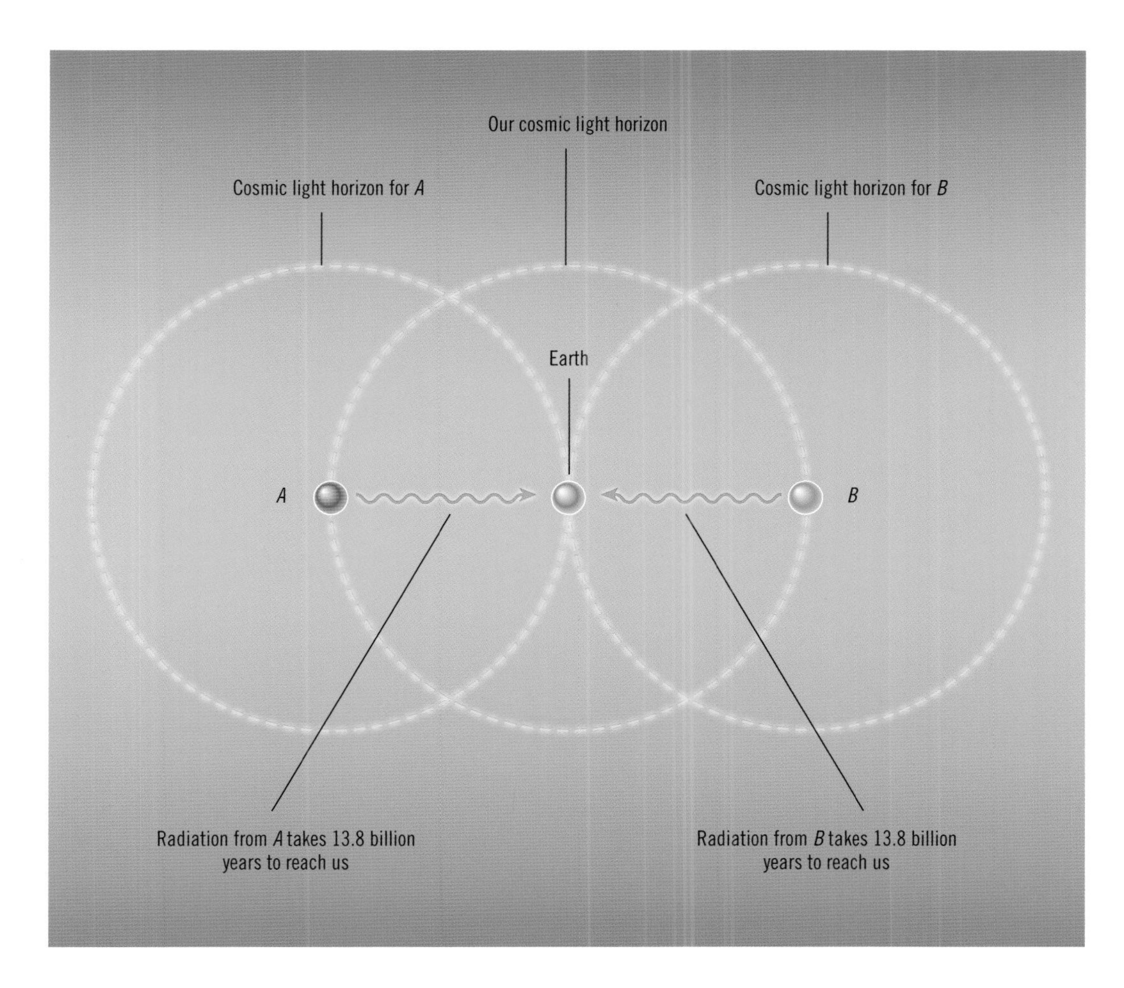

Above *After 13.8 billion years the light emitted by Objects A and B finally arrive at Earth. But although A and B can see Earth, they cannot see each other until the universe becomes 27 billion years old. Every object in the universe is at the center of its own visible universe horizon, which is a unique feature of Big Bang cosmology.*

fact, there would be numerous quantum black holes forming and evaporating, so the universe was a cauldron of energy and wildly-changing space geometry.

The expansion of the universe proceeded at an incredible rate, so that by about 10^{-36} seconds ABB the universe had grown in scale by some 3,000 times. The temperature had fallen to about 10^{29}°C, and the most massive particles created from vacuum fluctuations were about 10^{15} times the mass of a proton. Particles that used to be 10^{-35} meters apart were now 3,000 times further apart at 3×10^{-32} meters. The expansion of the universe according to the Big Bang model leads to a very important feature: our visible universe is only a small part of a larger unobservable space. This is called the cosmic horizon effect.

The size of your visible universe is limited by the distance that light could have traveled had it started out its journey at the instant of the Big Bang. This means that 10^{-36} seconds after the Big Bang, light could only have traveled a distance equal to 300,000,000 m/s x 10^{-36} seconds or 3×10^{-28} meters. If two particles, A and B, started out at 10^{-30} meters apart, by 10^{-36} seconds ABB they would be 3×10^{-27} meters apart. But light could only have traveled 3×10^{-28} meters, so A and B would be too far apart for light to travel between them and so each particle would have a visible universe, called a horizon, surrounding them that has a radius of 3×10^{-28} meters. Today, our visible universe has a radius of 13.8 billion light-years, but from the Big Bang there is still a lot of universe with galaxies and stars beyond this distance. Light from these objects is still on its way to us since it began its journey at the time of the Big Bang. If we wait another billion years, we will start to see the light from objects that are now 14.8 billion light-years from us according to their light-travel distance. Of course, the images we see will be of what these objects looked like 14.8 billion years ago.

During the universe's initial expansion period that ended about 10^{-36} seconds ABB, the furthest distance you could see was by that time 3×10^{-28} meters, and your visible universe was starting to get crowded, but the density of matter and energy in space was still an unimaginable 10^{85} kg/m3. Once again, we don't know what forms of matter existed at this time, only that, whatever was going on, the enormous vacuum fluctuations in the gravitational field of the universe were capable of spawning very massive particles if they could exist at all. This physics is beyond the Standard Model to describe in any detail, which has only been tested to temperatures of 10^{15}°C. Since the 1970s, however, physicists have worked on a way to unify all three forces in the Standard Model in a variety of theories called Grand Unification Theory (GUT). These make very similar predictions for what to expect at these cosmic energies.

GUT models predict that at 10^{29}°C the strong force will become similar in strength to the electromagnetic and weak forces. Also, all of the 25 elementary particles in the Standard Model will be completely massless, just like the familiar photons and gluons are today. The Standard Model will also be supplemented for the first time by a hypothetical new family of supermassive particles that physicists call the X and Y leptoquarks. These particles will each be about 10^{15} times more massive than a single proton. The vacuum fluctuations in "empty space" at 10^{-36} seconds ABB will be creating these particles and their antiparticles in huge numbers. At this time there will only be two kinds of distinct forces in the universe: gravity and the unified GUT force. Just as in the Standard Model where the Higgs boson gives mass to the other particles, in GUT world there is a supermassive Higgs boson that does the same thing for the strong force. This field is embedded everywhere in space, but vacuum fluctuations in this field spawn supermassive Higgs bosons that interact with the X and Y particles, giving them masses too. As the universe expands and continues to cool, everything is going well until an energy threshold is reached. It is much like the threshold in temperature that separates liquid water (high energy) from solid ice (low energy) at 32°F (0°C). It is a threshold that has a very fortunate consequence for us, but a catastrophic impact upon the entire universe.

Below *The growth of the universe from the Big Bang.*

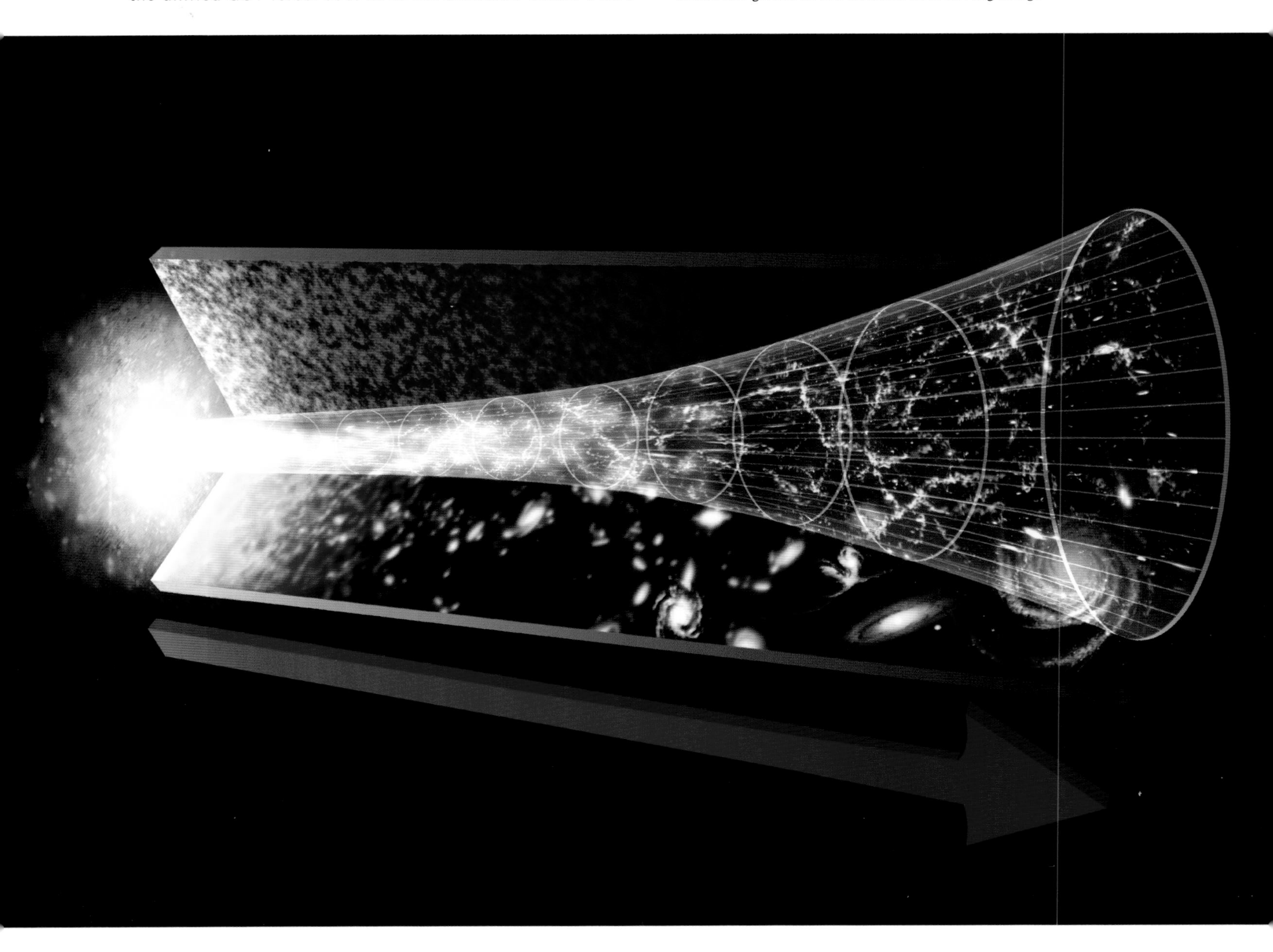

// Inflation and the observable universe

The time period from about 10^{-36} to 10^{-34} seconds ABB is one of the most critical eras in forming the present-day universe. The details for what happened during this period are described by what astronomers call inflationary cosmology, developed by physicists Alan Guth and Andre Linde in the early 1980s. It all starts with the supermassive Higgs field, or perhaps a cousin to it called the inflaton field. The energy of the Higgs/inflaton field is determined by how the Higgs/inflaton particles are interacting with each other, and this changes dramatically as the universe expands and cools. This field has two "minimum" energy states. The first minimum occurs at the very high temperatures that existed before 10^{-36} seconds ABB. But as the universe expands and cools, some regions of this field find themselves in a lower energy state that also happens to be a minimum energy much like the bottom of a rollercoaster ride between its two peaks.

To make the transition from the higher to the lower energy state, the Higgs/inflaton field has to go through a quantum tunneling event, and this causes the universe to exponentially grow in scale. Prior to 10^{-36} seconds ABB, the universe doubled in scale every time the universe got twice as old. But during this inflationary expansion, the universe doubled in size about every 10^{-36} seconds, so between 10^{-36} and 10^{-34} seconds ABB it underwent 100 doublings in scale.

Below *Bubbles forming in a glass of champagne provides an analogy for the formation of numerous bubble universes out of the rapidly-inflating spacetime of the early universe. Our visible universe would be the size of a dust speck inside one of these champagne bubbles.*

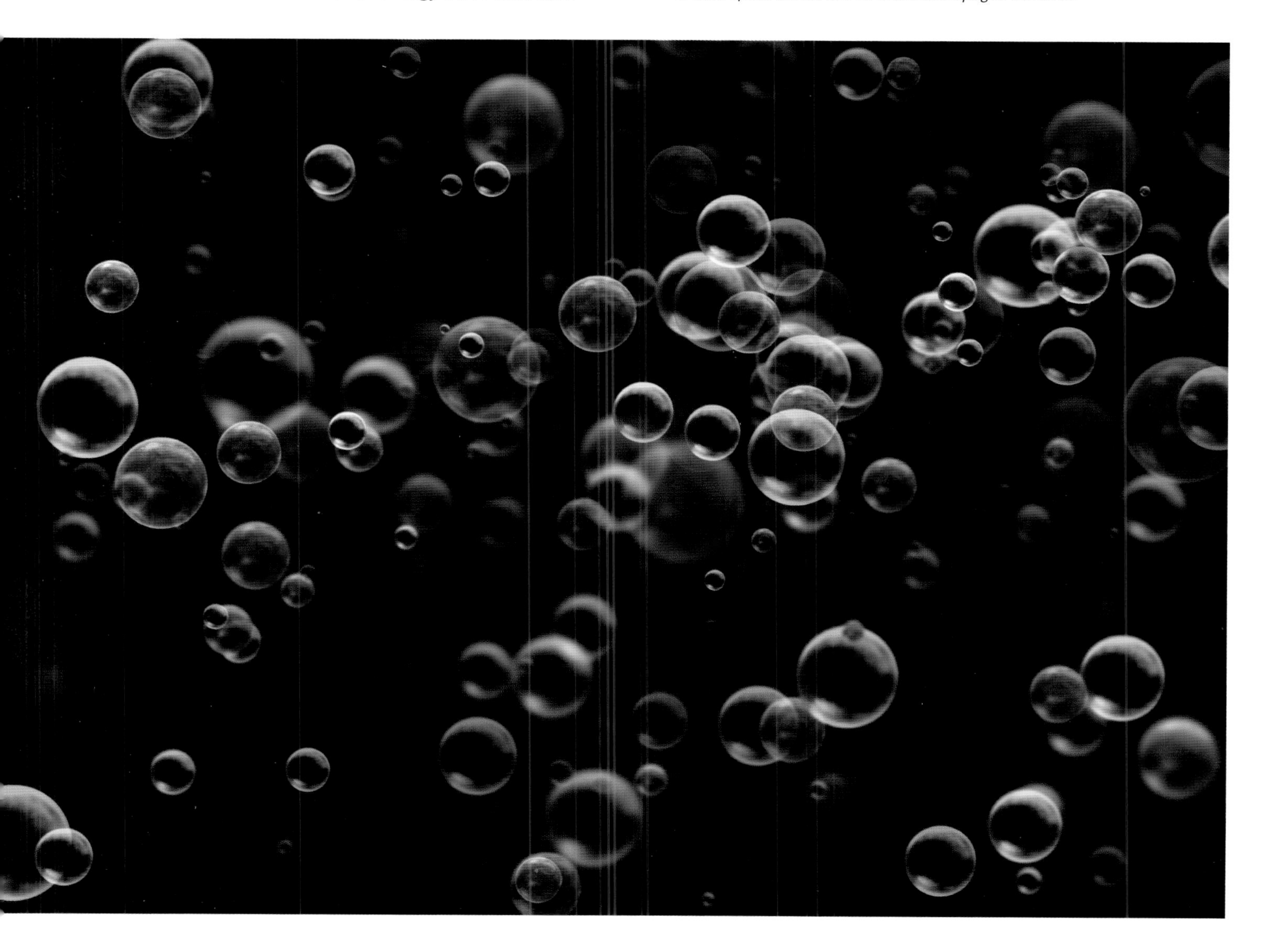

This would have dramatically cooled the universe to almost absolute zero except that, at the end of the tunneling process when the field finally reached its new minimum energy, it re-heated the universe to nearly the temperature it had at 10^{-36} seconds ABB, unleashing a firestorm of GUT and supermassive Higgs bosons flooding into existence in space. At this time, some 10^{-34} seconds ABB, instead of the scale of the universe being only about 10^{-31} meters, the 100-doublings had dilated space to a scale of $10^{-31} \times 2^{100} = 0.1$ metre, which is about the size of a grapefruit. This entire space was now a seething cauldron of GUT particles at a temperature that was still close to 10^{29}°C.

The transition from the higher vacuum, called the false vacuum, to the lower vacuum, called the true vacuum, is akin to the formation of bubbles in a liquid. Each bubble is a region of true vacuum that is expanding like the regular Big Bang theory predicts, but in between the bubbles we have the false vacuum, which is still trying to tunnel into the true vacuum. This means that in between

Above *This 3D rendering shows the changes in density for a universe that has expanded 200-fold since the end of the inflation era. Brighter colors indicate higher densities.*

the bubbles, one of which grows to become our universe, we have an exponentially expanding space that is very rapidly separating the bubbles to enormous distances. Also, the particles produced inside these true vacuum bubbles are not smoothly spread out in the available space, but are clumped according to various quantum rules. By the present age of the universe 13.8 billion years later, inflationary cosmology predicts we can still see the vestiges of this inflationary quantum lumpiness within our own small corner of our particular cosmic bubble. We see it in the way the largest structures in the universe such as galaxies and clusters of galaxies are spread out in space. Amazingly, the quantum physics of the Big Bang left a cosmos-spanning mark in the way our universe looks today at vastly larger scales.

// The fireball radiation

During the inflationary era when our universe was exponentially expanding in scale and settling down into the true vacuum, GUT predicts that empty space would have been flooded with a host of X and Y leptoquarks and their antiparticles such that for every leptoquark, there was exactly one antileptoquark. For a time, as soon as one of these pairs was created out of the supermassive Higgs/inflaton field, it would be annihilated and so the universe during this time was a delicate balance between creation and destruction. But as the universe continued to expand and cool, there was no longer any energy available to keep forging these particle-anti-particle pairs out of the vacuum of space.

Without any other assistance, the balance would be tipped to total annihilation and so all of the matter and antimatter leptoquarks in the universe would vanish completely. They would be replaced by pairs of photons of light, each carrying as much energy as an individual X or Y particle. The universe would have evolved from that time forward as a gas of photons with no matter left over to form atoms, stars, and galaxies. This is clearly NOT the universe we live in, so something else must have happened to change the matter-antimatter balance.

Today, we can measure the number of photons in the cosmic background radiation (astronomers call this the CBR) that was left over from the Big Bang and compare this to the number of quarks in the universe that have built the stars and galaxies around us. The answer is that there are about 10 billion photons for every quark in our neck of the universe, and presumably everywhere else, too. What this means is that just after inflation ended, instead of there being 5 billion particles and 5 billion antiparticles to annihilate into 10 billion photons and zero leftover matter particles, there were 5 billion-and-one matter particles and 5 billion antiparticles, which would annihilate to give 10 billion photons and 1 matter particle. Physicists have no good ideas how such a minuscule imbalance of 1-part-in-10 billion could have been achieved by either Standard Model physics or the extended physics offered by GUT, but finding an answer to this antimatter mystery is one of the most pressing avenues of research in modern astrophysics today. But amazingly enough, our Big Bang story continues despite resolving this conundrum. Our observations of the universe today only have to do with the leftover matter and how it evolved in time.

Below *Although matter and anti-matter represent the two sides of a coin of matter, for reasons not yet understood, the universe "flipped" this coin so that for every 10 billion particles of anti-matter there would be 10 billion-and-one particles of matter.*

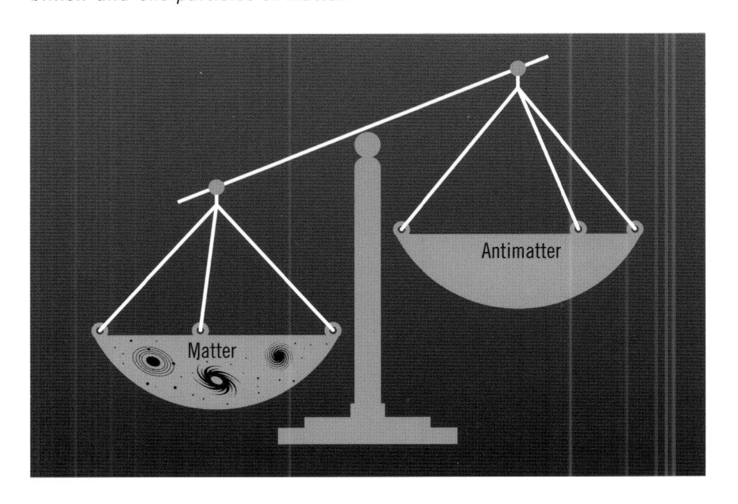

Once inflation had ended and we reached the condition where space was filled by the CBR and a trace of matter, the expansion continued to cool the universe even further until the X and Y leptoquark matter particles began to decay, but this time in a colder universe where they could no longer be replenished. These massive particles decayed into the quarks and leptons of the Standard Model, but at these temperatures the Higgs boson of the Standard Model was massless and so all of the quarks and leptons of the Standard Model were massless too, and traveling through space at essentially the speed of light. This takes care of the Standard Model matter but we still do not know how dark matter made its appearance in this story because we have no good ideas what kinds of particles they may be. All we do know is that they interact with Standard Model matter through their gravity and by no other means. The genesis of dark matter is the second missing chapter in the modern cosmological story, but again the events that happened to matter after inflation and for the next million years have nothing to do with dark matter because it only interacts with normal matter through very weak gravitational forces. So far as we know, dark matter forces are not strong enough to upset any of the nuclear physics taking place during the first ten minutes after the Big Bang.

Following the end of inflation at 10^{-34} seconds ABB, GUT predicts the very sad story that no new forms of particles enter the picture beyond the known particles in the Standard Model. Physicists call this the Particle Desert. But if string theory is found to be correct, it does fill in this period of

Right *Conceptual illustration depicting the Big Bang and yet-undiscovered dark matter particles. Dark matter is assumed to have formed after the Big Bang.*

time with new populations of particles. Meanwhile, there are a number of other distinct eras that come and go involving the Standard Model particles as the universe continues to expand and cool. The most important of these is the time when the weak and electromagnetic forces finally became distinct forces. From the Standard Model, this would happen at a temperature of about 1,000 trillion°C, and Big Bang cosmology predicts this temperature occurred when the universe was about 10^{-10} seconds old.

Current technology such as the Large Hadron Collider at CERN and the Relativistic Heavy Ion Accelerator (RHIC) at Brookhaven allow us to explore energy conditions that existed in the universe between 10^{-13} seconds ABB and 10^{-6} seconds ABB. By recreating the kinds of energetic col-lisions that occurred at that time, physicists can study how the Standard Model particles interacted. One of the most recent discoveries is of a new state of matter called the quark-gluon plasma. This existed when quarks and gluons behaved like particles in a gas during the time just before the quarks assembled themselves into the familiar protons and neutrons. This era ended once the universe cooled to below about 10 trillion°C at a time of about 10^{-6} seconds ABB.

Right *The Large Hadron Collider is a gargantuan particle accelerator 17 miles (27 kilometers) in circumference where protons are slammed together at nearly the speed of light to discover new elementary particles and to test the Standard Model of physics.*

Below *A computer visualization of a quark-gluon plasma in the Relativistic Heavy Ion Accelerator (RHIC) collider at the Brookhaven National Laboratory, which represents the form of matter in our universe when it was only one millionth of a second old.*

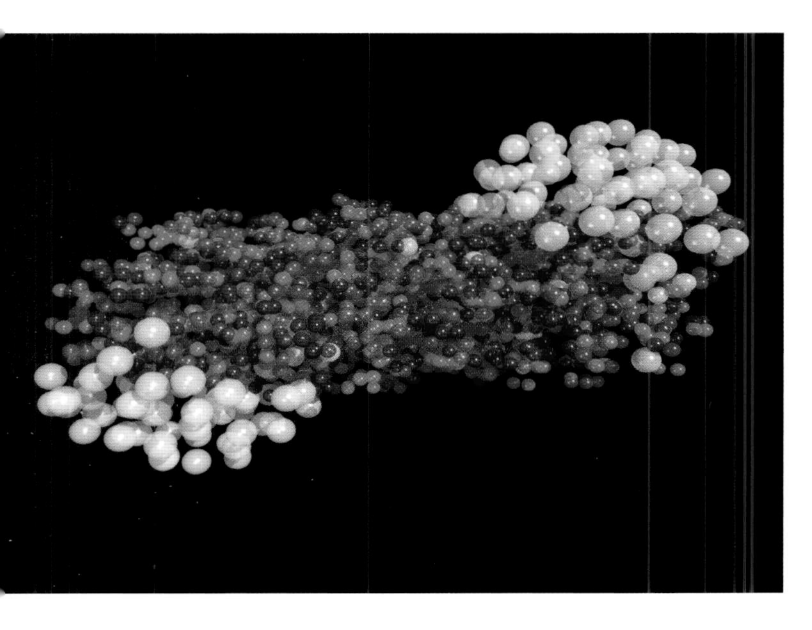

// Nucleosynthesis era

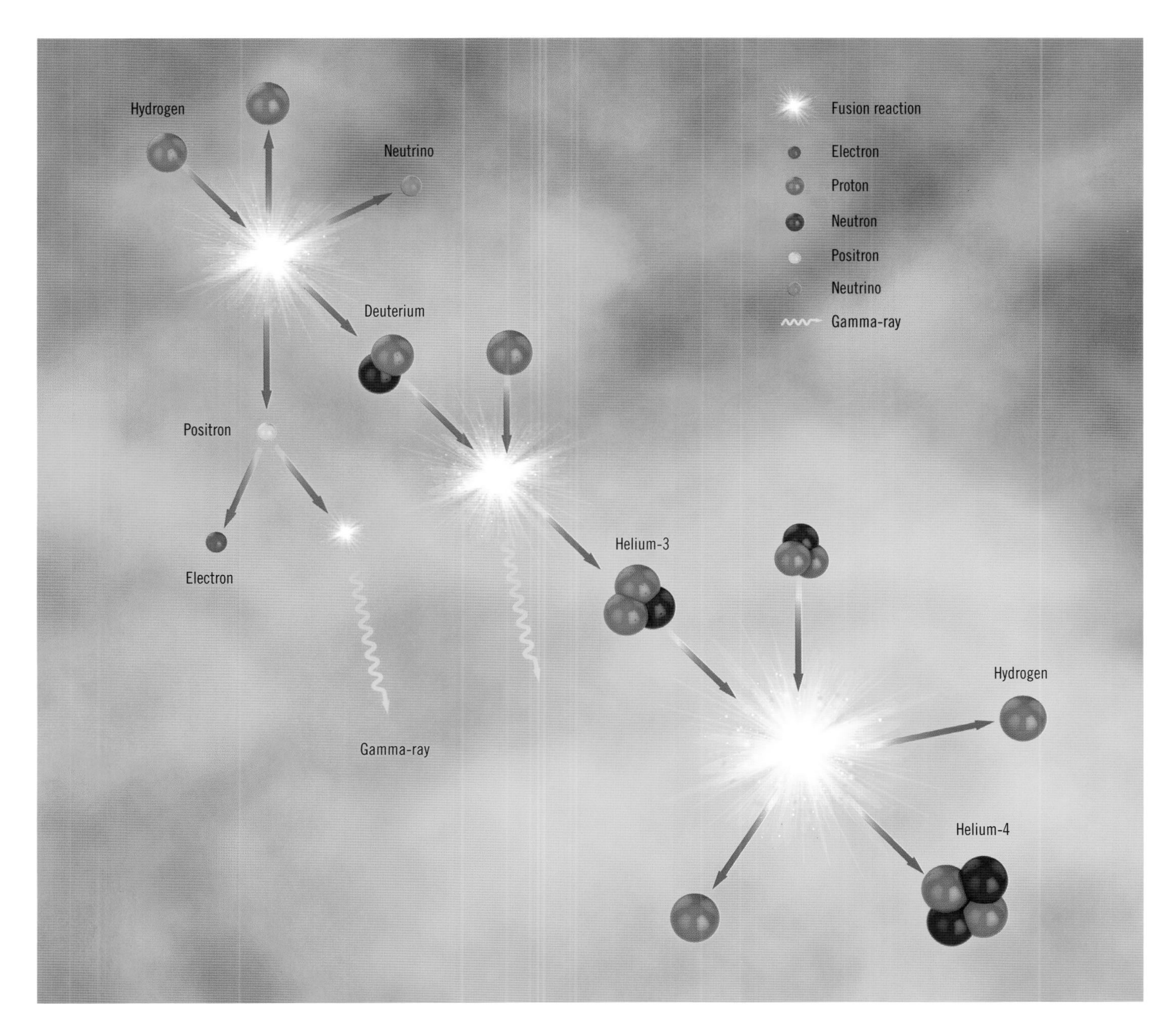

Above: *Two hydrogen nuclei fuse to form a deuterium nucleus, a positron, and a neutrino. The positron quickly encounters an electron, they annihilate each other, and only energy remains. The deuterium nucleus goes on to fuse with another hydrogen nucleus to form helium-3. In the final step, two helium-3 nuclei fuse to form helium-4 and two hydrogen nuclei.*

The next significant era was the era of nucleosynthesis that began about one second ABB. The temperature of the CBR was about 10 billion°C, and our universe had relentlessly expanded to enormous dimensions since the end of the inflation era. The bubble that would contain our universe, which had emerged at the end of the inflationary era, was at that time about 1 meter across, but by the current time of one second ABB it is now 10^{17} times larger or about three light-years. But because by this time the universe is only one second old, the furthest distance you could see would have been only one light second. In a scale model, if you were to draw a circle the radius of Earth to represent the scale of our bubble at this time, the light horizon distance would be only the diameter of an apple. Inside this Earth-sized model, the temperature and contents of the universe would be the same—the same

Standard Model; the same physical laws; the same physical constants. What we are able to see in our own little apple-sized corner of this vaster universe is only a very small part of a larger universe bubble that emerged from the Big Bang. Today the diameter of this bubble has swelled to over 1,000 trillion light-years, but we can only see the contents of our observable universe out to about 13.8 billion light-years.

The nucleosynthesis era is a key step in creating the contents of our observable universe because it is during this time that the primordial elements hydrogen, helium, lithium and deuterium were created from the raw protons and neutrons that were available. But this creation process by collisions is running a race that it can never win. To fuse particles together requires high temperature and density, but the expansion of the universe is relentlessly causing both of these properties to fall.

At two minutes ABB, the density of the universe was about 10^{-4} kg/m^3 and the temperature of matter was about 1 billion°C. By the end of this era some 20 minutes ABB, the temperature was only 300 million°C and the density had fallen to 10^{-10} kg/m^3. For comparison, the density of the protons, neutrons, and electrons in the plasma that fills all of cosmic space is already about like that of Earth's atmosphere at an altitude of 43 miles (70 kilometers). By 20 minutes ABB, the density is more like our atmosphere at an altitude of 124 miles (200 kilometers) just below the orbit of the International Space Station.

By the time the nucleosynthesis era started, the ratio of protons to neutrons was about one neutron for every five protons. Many of the free protons (the nuclei of hydrogen atoms) would be able to bind with available neutrons to form deuterium nuclei. At first it was a fragile bonding because there were enough free particles around to blast these deuterons apart, but as the universe continued to cool, the initial balance between formation and destruction shifted to deuterium nuclei persisting. At the same time, the deuterium nuclei collided with the ambient neutrons and protons to form tritium nuclei (an isotope of helium with two protons and one neutron) and then helium nuclei (two protons and two neutrons). There was even some ability for the helium nuclei to gain protons and neutrons to become lithium (3p, 4n) and beryllium (4p, 3n) nuclei. In fact, beryllium formed when one of the lithium neutrons decayed into a proton.

The struggle to form nuclei more massive than beryllium came to an abrupt end because the universe was now too cold for the free protons to overcome the repulsion of the lithium or beryllium nuclei. The build-up of all of the other elements in the periodic table, including carbon and oxygen necessary for life, had to wait for other opportunities to occur in the evolving universe. By 20 minutes ABB, the formation of more of these primordial elements from hydrogen to beryllium came to an end. Any remaining free neutrons not located by this time inside atomic nuclei rapidly vanished. A free neutron left by itself for about 880 seconds will decay into a proton, an electron, and a neutrino. By the end of the first year ABB, the freezing-in of several critical cosmic ratios occurred that can only be understood by using Big Bang cosmology, making this another unique and successful prediction by the theory. On the cosmic scale, primordial matter should consist of 24 percent helium; 76 percent hydrogen and traces of deuterium and lithium far below 1 percent in abundance. This, then, is the bedrock matter from which all familiar stars and galaxies are created. In the CBR, there are still ten billion photons for every quark, and in addition thanks to all of the decays and other interactions involving the weak force, there is a CBR of neutrinos flowing through the universe equal to about one neutrino for every four CBR photons.

Below *Big Bang theory predicts the abundances of light atoms such as helium, deuterium, and lithium. These abundances (vertical axis) can be measured today and depend on the density of matter (horizontal axis), which can be used to test the accuracy of Big Bang theory.*

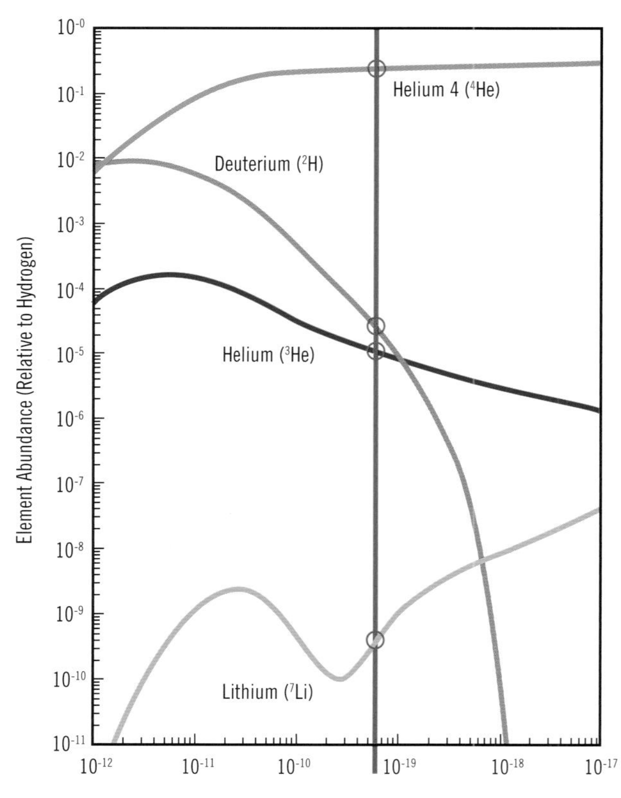

// The cosmic background radiation era ends

The end of the nucleosynthesis era heralded the beginning of a very long cosmic episode when not much happened. The universe continued to expand and the temperature of matter relentlessly cooled as the universe became less and less dense. Because matter particles were still charged, they continued to interact strongly with the CBR, making the universe at this time resemble a Pachinko game with the CBR photons bouncing between the particles. The CBR photons collided with protons, atomic nuclei, and the numerous free electrons and scattered into all possible directions, losing their energy. At the same time, the matter particles collided with the photons and boosted their energies. This process remained in equilibrium so that the matter and photons acted like a gas. But for most of this time, the photons had the upper hand.

There are two kinds of gas pressure in the early universe; the pressure from the CBR photons, and the pressure from the matter particles. For all of the history of the universe after inflation, it has been the pressure from the CBR radiation that has determined how fast the universe would be expanding. But a gas of photons behaves slightly differently than a gas of matter particles. As the volume of space increases and the expansion of space causes wavelengths to get stretched, the CBR pressure falls more rapidly than the matter pressure. At a time of about 16,000 years after the Big Bang these pressures become equal, and thereafter the matter pressure starts to be more important. Now it is the pressure of matter particles that drives the expansion of the universe. At this time, the temperature of the universe has fallen to about 1 million°C. The universe is still a plasma of nuclei and photons at a density of about 0.001 kg/m^3, and photons and nuclei are still colliding, but the CBR photons are rapidly losing their control over matter.

By the time the universe has cooled to about 3,000 K, electrons are able to be attracted to protons and helium nuclei to make them neutral atoms for the first time in cosmic history, but the CBR photons no longer have enough energy to blast them apart into a plasma state again. At this point, the universe becomes transparent to the CBR light, causing the neutral matter and the CBR photons to go their own ways in cosmic history from now on. This critical event in the evolution of matter in the cosmos is called the recombination era and it marks a moment in history dating from about 380,000 years ABB.

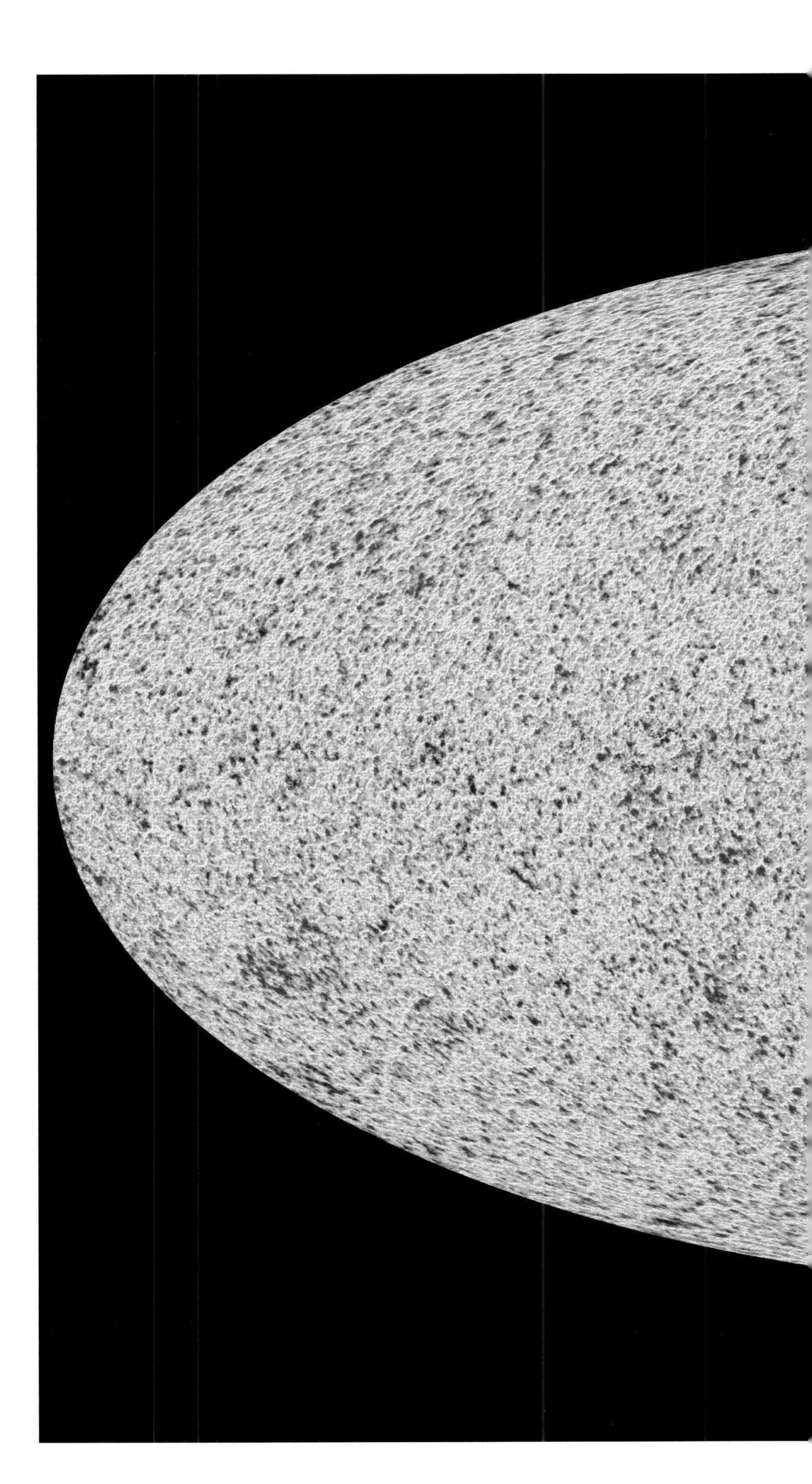

Below *The Planck spacecraft map of the cosmic microwave background reveals irregularities at a time about 380,000 years ABB, which later led to the formation of clusters of galaxies in the modern universe.*

// Dark matter structure

At the end of the recombination era 380,000 years ABB, the temperature of matter is about 5,432°F (3,000°C). If you had been there, in every direction you looked you would see the "empty space" in the universe glowing like the surface of a dull red star. But this condition would not last long. Again, as the universe is expanding and cooling under the relentless action of matter pressure, by the time you reach an age of 2 million years ABB, the cosmic glow has vanished to human eyes and is now starting to rapidly recede, first into the infrared spectrum, then into the far-infrared spectrum. Meanwhile, the primordial hydrogen and helium gases are so cold they would be invisible to the human eye at optical wavelengths. Astronomers call this the start of the Cosmic Dark Ages, but it is far from being a dull time in cosmic history. It is at this time that dark matter begins to gain its upper hand.

Dark matter has been with us as a silent partner since the inflationary era, but has had an invisible hand in

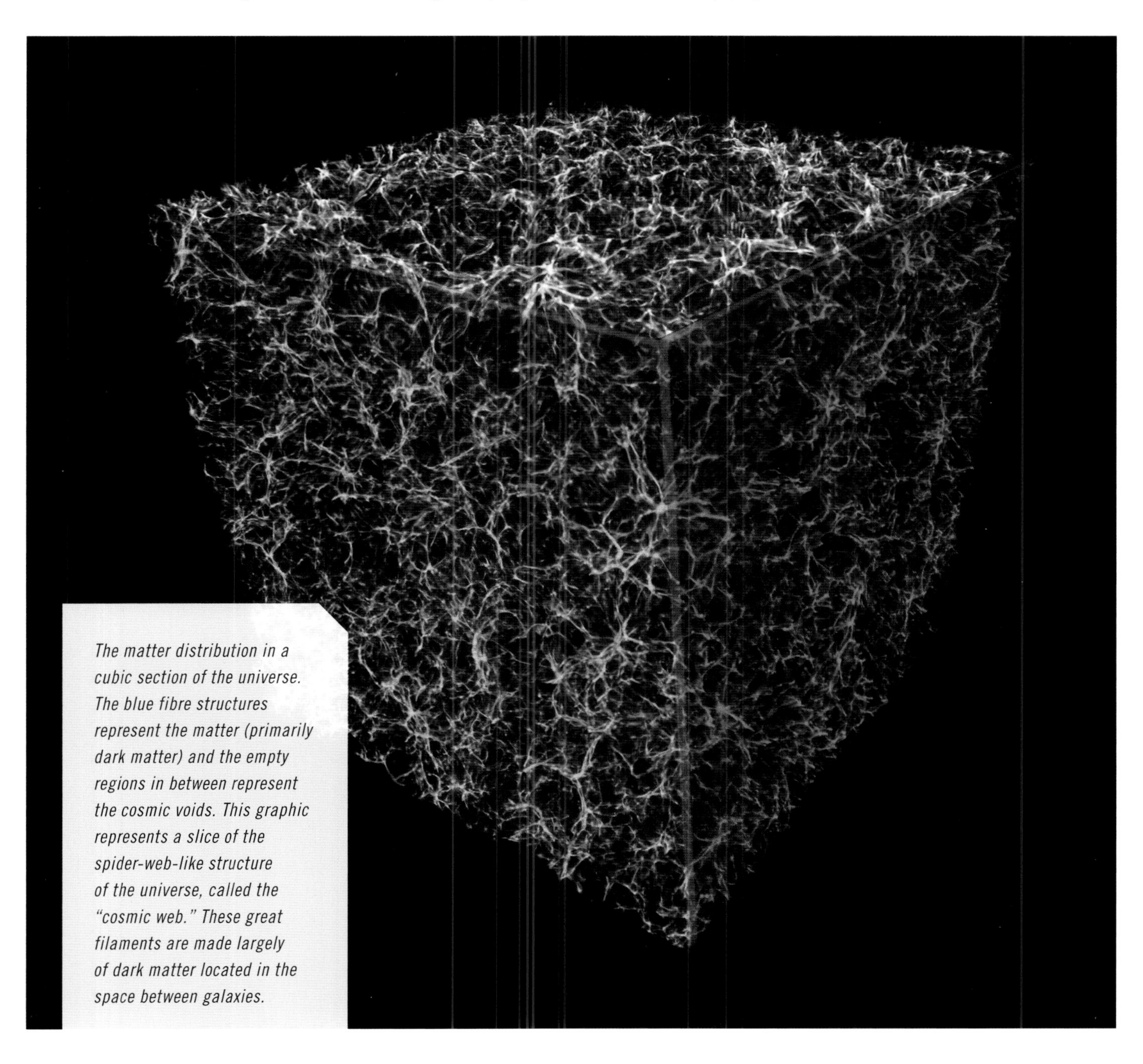

The matter distribution in a cubic section of the universe. The blue fibre structures represent the matter (primarily dark matter) and the empty regions in between represent the cosmic voids. This graphic represents a slice of the spider-web-like structure of the universe, called the "cosmic web." These great filaments are made largely of dark matter located in the space between galaxies.

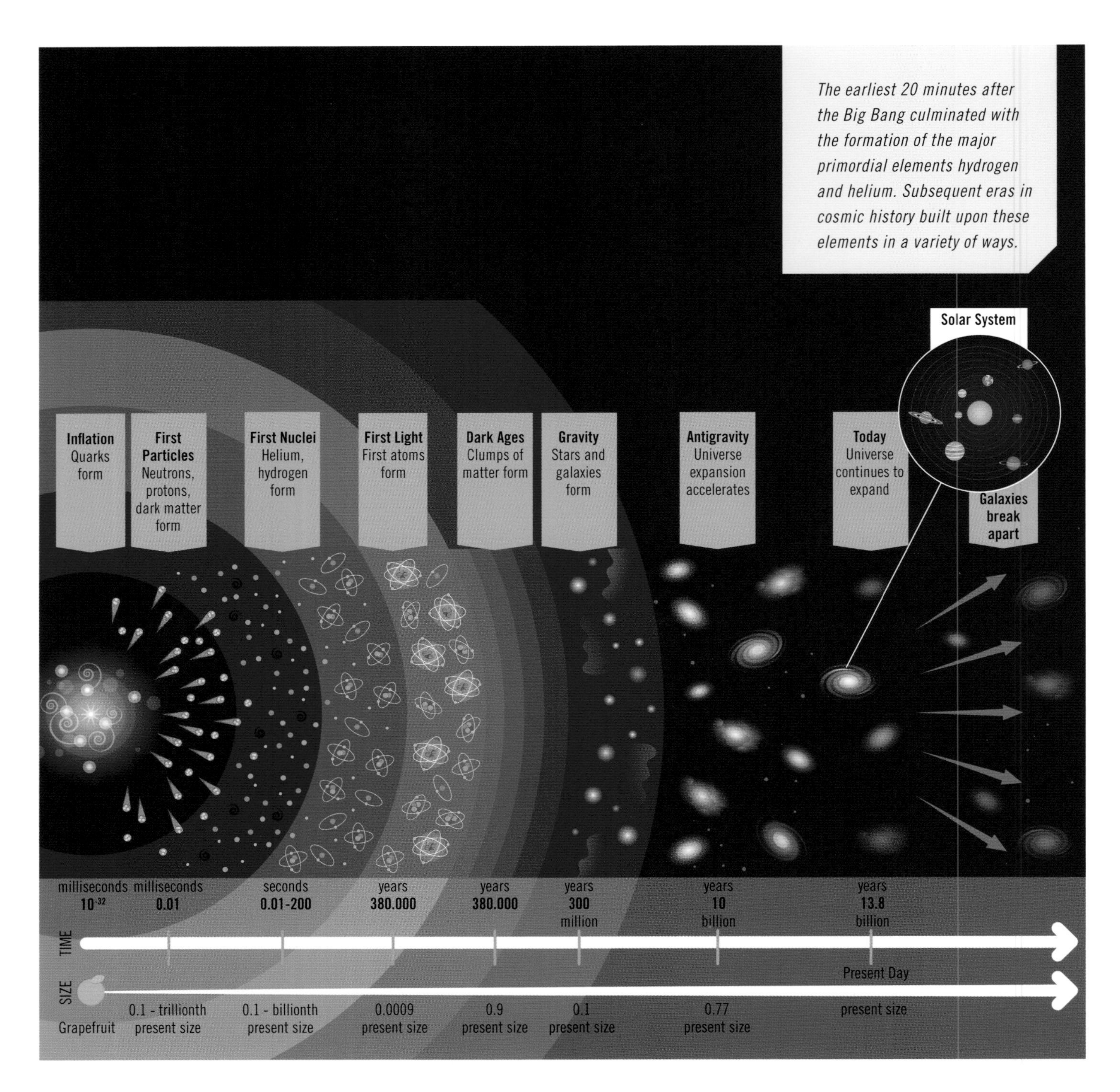

The earliest 20 minutes after the Big Bang culminated with the formation of the major primordial elements hydrogen and helium. Subsequent eras in cosmic history built upon these elements in a variety of ways.

how the universe looks. It interacted only by virtue of its gravity, but only now has it begun to have an effect on matter. Dark matter is known to be lumpy on large scales in the universe. Instead of space filled with a uniform gravitating material it is a complex patina of dark matter gravitational wells, and other structures. Over time, supercomputer models show that dark matter formed itself into vast ribbons and gravitational pits stretching for millions of light-years. By the time the Dark Ages were well in hand, matter had become cool enough to begin to fall into these dark matter gravity wells forming vast clouds of matter imprinted with the dark matter structures. The stage was now set for the rapidly cooling normal matter in these wells to collapse under their own gravity into still-smaller scales to form the first stars and galaxies. To understand this era a bit better, let's have a look at how stars, the building blocks of structure in our universe, form and evolve.

STELLAR EVOLUTION

Stars are among the most important objects in the universe. They provide light and energy to warm the surfaces of planets, but they also provide the beacons that astronomers use to map out the landscape of the universe. The earliest stars to form in the universe were dramatically different than the ones we know today. Their deaths as supernovae created all of the elements out of which planets and living systems would eventually form. With powerful telescopes on the ground and in space, astronomers hunt the outskirts of our visible universe to catch glimpses of these first stars shining dimly across 13 billion years of cosmic history.

The night sky can be photographed with nothing more than a smartphone camera and reveals the dazzling scope and breadth of the cosmos as perceived from the human standpoint in space and time. The color and brightness of the stars we see are glimpses of the evolution of these clumps of matter across millions of years.

// Observing the stars

Since before the dawn of human history, stars have lit up the night-time skies causing our ancestors to wonder what they were. Unlike the familiar trees, mountains, and rivers of the terrestrial world, the night-time sky is the most alien vista anyone could possibly comprehend in terms of familiar objects. During recorded history beginning 7,000 years ago, we see how our ancestors developed mythologies to help explain what they were seeing. Generally, stars were considered to be living celestial beings that formed constellations whose shapes were appropriate to each deity. By the time of the Ancient Greeks and Aristotle *c.*300 BCE, stars and planets were considered to be fashioned out of a fifth element called "aether," and later called "quintessence" by medieval alchemists. It was a luminous substance that could only move in perfect circular motion. It would take until the 16th century before stars were considered to be suns in space, and until the 19th century before their elemental compositions could be detected using a new instrument called the spectroscope. Stars were now just familiar kinds of matter but seen under unfamiliar circumstances.

The distances to the stars was a matter of speculation for millennia until the German astronomer Friedrich Bessel used the parallax method in 1638 to estimate that the bright star 61 Cygni was at a distance of nearly 62 trillion miles (100 trillion kilometers/10.4 light-years) from Earth. Over the next century, the distances and luminosities of thousands of additional stars were measured. Once astronomers were able to calculate the masses, temperatures, and luminosities for stars it became pretty obvious that stars came in a wide range of sizes and temperatures, from diminutive white dwarfs such as Sirius B, only about the size of Earth, to vast red supergiant stars such as VY Canis Majoris, capable of engulfing our entire solar system to the orbit of Saturn.

The idea that stars were born and evolved over time did not actually come into existence until the mid-1800s, perhaps even inspired by Charles Darwin's seminal works in evolutionary biology. Instead, the origin of stars was tied up with the biblical idea that the universe was only 6,000 years old and formed as we see it for the rest of eternity. But physicists such as Hermann von Helmholtz and Sir William Thomson (Lord Kelvin) took a more pragmatic view. The concern was the ultimate source of energy for lighting the sun itself. If it was slow gravitational contraction, then the sun could be more than 100 million years old, placing it in the same timescale as Charles Lyell's studies of the formation of mountains on Earth. A number of stellar evolution ideas surfaced in the early 1900s, but it would take the discovery of thermonuclear fusion in the mid-1930s by physicists such as George Gamow to finally identify the fusion of hydrogen into helium as the main source of energy for stars like our sun. This led to a refined understanding of how stars evolved over time and how many of the individual types of stars were related to each other via evolutionary sequences. For example, over the course of the next eight billion years, our sun will deplete its core hydrogen reserves and swell into a red giant star, and from there lose its outer layers to become a white dwarf star surrounded by a planetary nebula. But before we jump into the details of stellar evolution, let us take a closer look at our nearest star: the sun.

Above *Friedrich Bessel was the first astronomer to estimate the distances to the stars using the parallax method. A similar technique has been used for centuries by surveyors to determine distances along the surface of Earth.*

Opposite *Artist rendering of the red supergiant star VY Canis Majoris. It is a variable star indicating it is unstable and likely emitting copious amounts of matter into space. With a mass equal to 25-60 times our sun, and with an age of 8 million years, it is soon to become a supernova.*

// Our sun as a star

The basic properties of our sun, a typical star, are nothing less than spectacular and even incomprehensible by human-scale dimensions. From a distance of 93 million miles (150 million kilometers), its radiant energy of 383 trillion trillion watts is enough to provide us with ample light and heat. It is over 100 times the diameter of our Earth. Within its bulk it can accommodate over one million planets similar to Earth. Jupiter is the only planet that comes close to rivaling the sun in size. Even so, the sun is 75 times more massive, and ten times larger in physical size. The sun accounts for 99.8 percent of the mass of the entire solar system. Its enormous gravity influences the motions of planets, asteroids, and comets to a distance of nearly two light-years; halfway to the nearest star system of Alpha

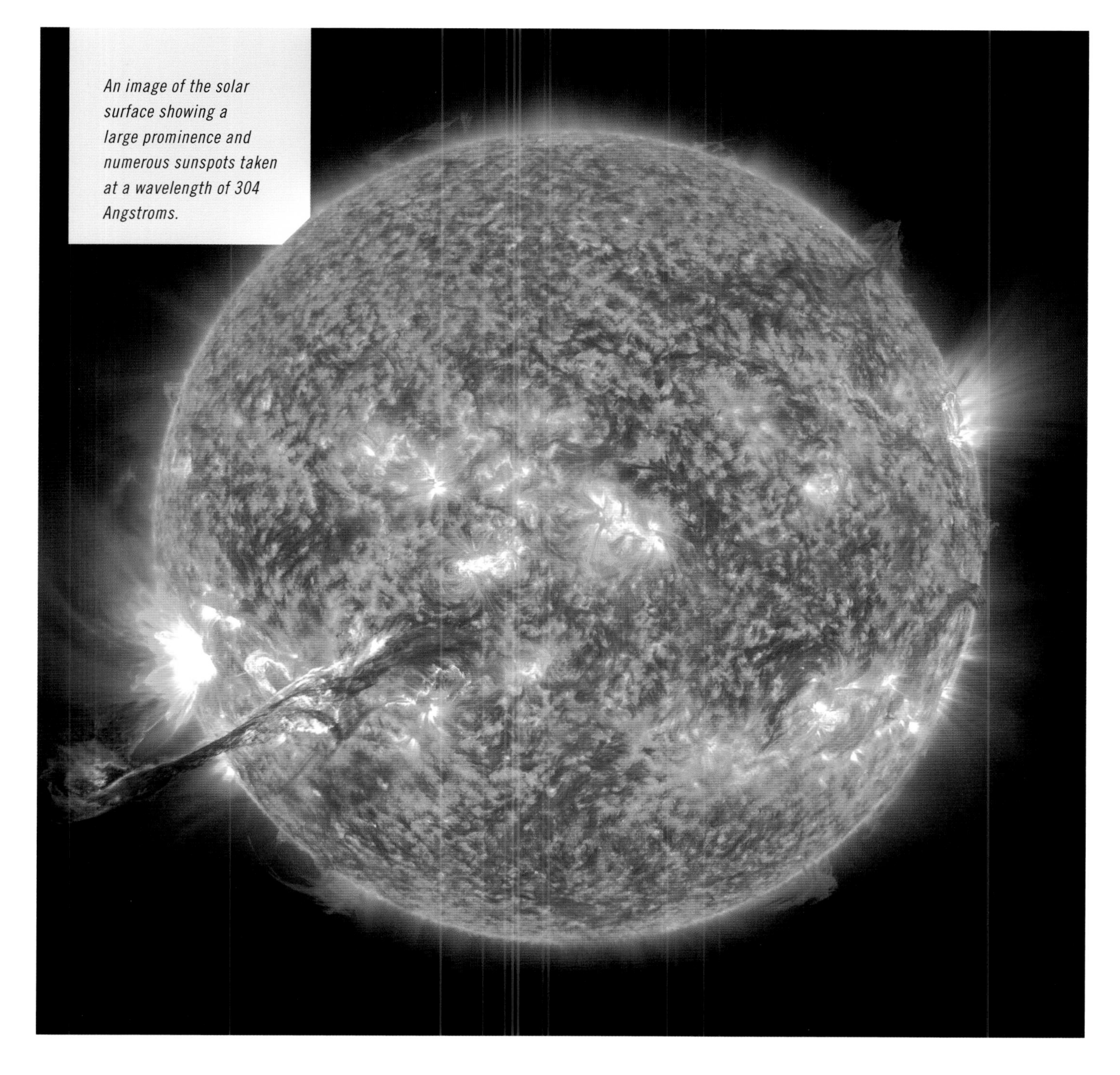

An image of the solar surface showing a large prominence and numerous sunspots taken at a wavelength of 304 Angstroms.

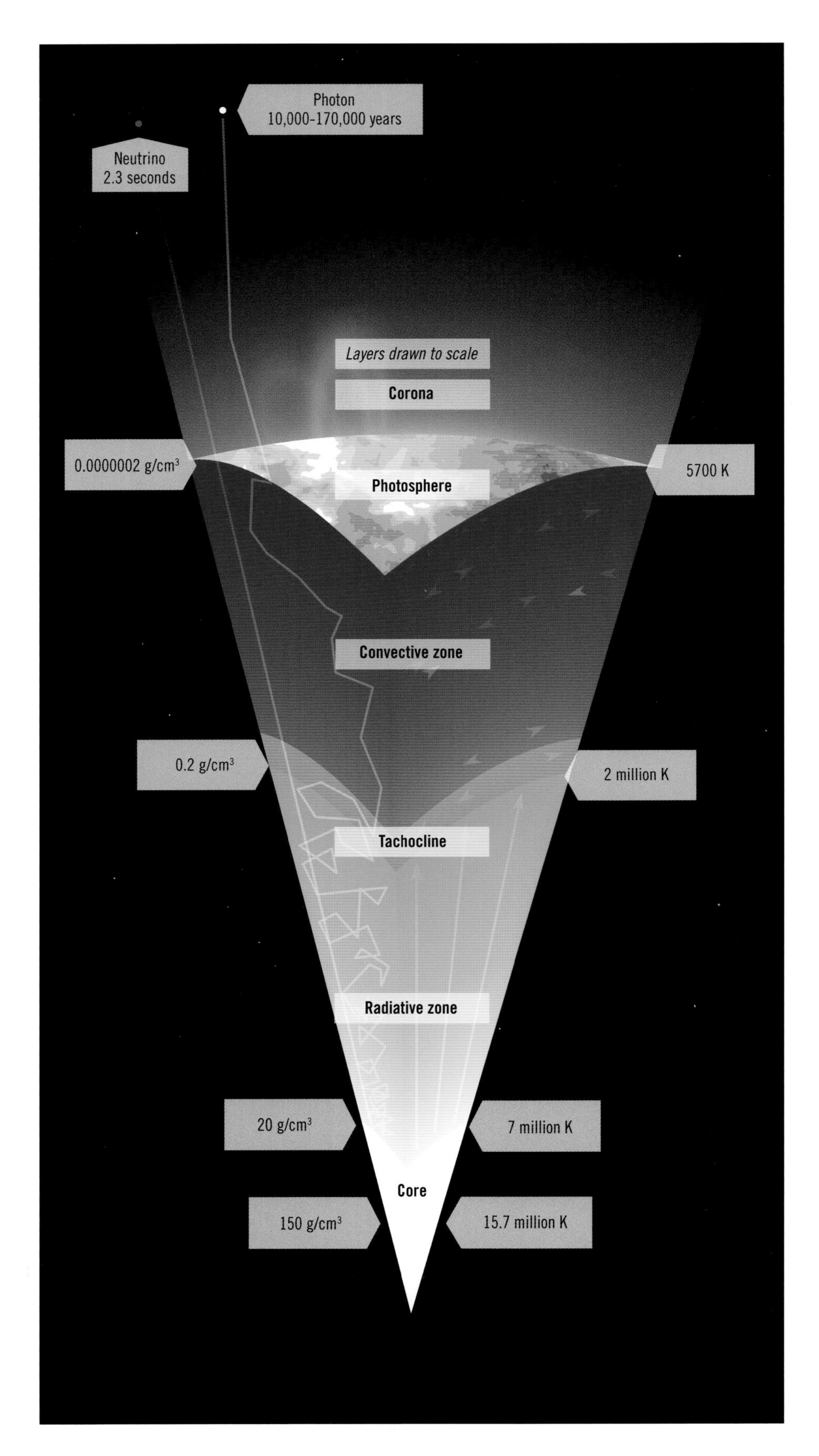

Centauri. Its visible surface, called the photosphere, is at a temperature of 9,932°F (5,500°C), which gives the photosphere its characteristic yellow color. The surface is composed of atoms of hydrogen and helium that have lost their electrons, which physicists call a plasma. Because of its strong magnetic field, this heated plasma forms sunspots, surface convection cells, and a very hot corona extending millions of miles into space. Astronomers have been studying the solar surface for centuries, but the real challenge is to understand our sun's interior, which is beyond direct observation. Since the early 1900s, physicists have created models of its interior based on a knowledge of how matter behaves at different temperatures and densities under the weight of its own gravity, beginning at the thermonuclear core of the sun.

As the amount of matter in an object increases, it stops being able to support itself through inter-atomic forces like a table or a planet. The in-fall of matter causes the center to heat up relentlessly until the constituent atoms and nuclei collide with enough energy to fuse into heavier nuclei. Our sun is 76 percent hydrogen, so the first step is to fuse hydrogen into helium, which releases "nuclear" energy. As the in-fall due to gravity increases, the nuclear energy produced creates a pressure that makes it harder and harder for more matter to accumulate. Finally, the gravitational force causing the object to collapse and heat up is balanced by the heat pressure created by the nuclear energy. This equilibrium causes the star to stop contracting so that it becomes very stable. If gravity tries to compress the star even slightly, the rate of nuclear energy production increases, pushing the gases outwards.

Left *A slice through the interior of the sun reveals many different processes.*

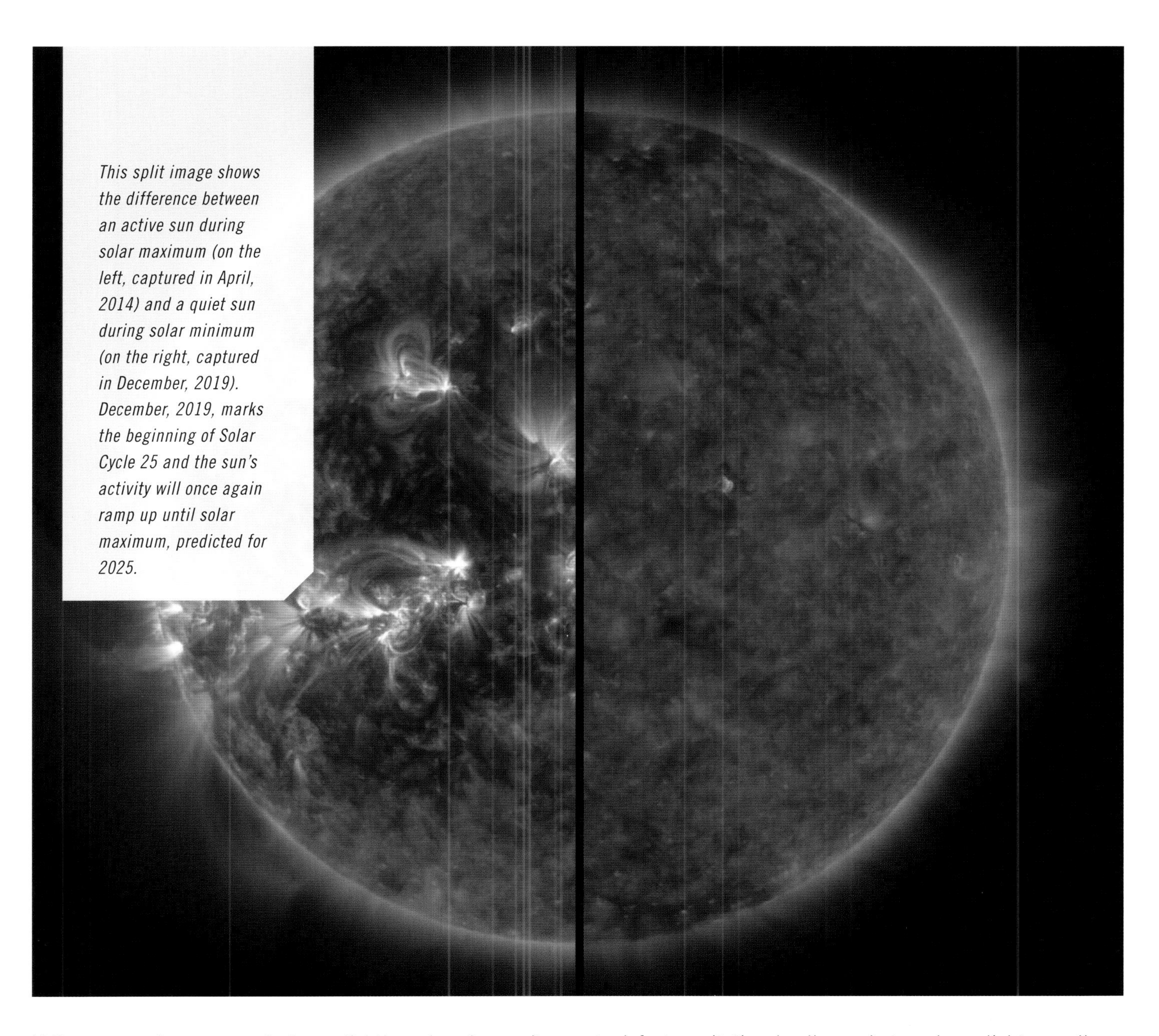

This split image shows the difference between an active sun during solar maximum (on the left, captured in April, 2014) and a quiet sun during solar minimum (on the right, captured in December, 2019). December, 2019, marks the beginning of Solar Cycle 25 and the sun's activity will once again ramp up until solar maximum, predicted for 2025.

If the core nuclear energy slackens slightly and cools, gravity wins the upper hand and compresses the star until the temperature and energy production returns to equilibrium levels. For a typical star, this equilibrium can be maintained for billions of years. The core of our sun in this equilibrium condition has a temperature of 15 million°C and a density of about 150 g/cc, or nearly ten times that of lead. Because this core region, the central 20 percent of the sun's radius, has been fusing hydrogen into helium for over four billion years, it has depleted its original 76 percent hydrogen to only about 34 percent.

The nuclear energy production in the core not only produces energy for maintaining the pressure inside the star to defeat gravitational collapse, but produces light as well. About 672 million tons (610 million tonnes) of hydrogen are converted into helium every second, liberating enough fusion energy to light the sun and provide its internal kinetic pressure. The light energy streams out from the core of the sun to reach its surface and then escape into space. This takes about 170,000 years for the newly-minted photons in the core to diffuse outwards through the interior of the sun. The collisions of the original gamma-ray photons with numerous atomic nuclei spawn millions of lower-energy photons that eventually reach the surface, and in 8.5 minutes pass Earth's orbit. Along the way, this transport of radiant energy alters the interior of the sun in a very specific

way. For most of the interior overlaying the core, this radiant energy is the main source of energy transport rather than the movement of matter, so this region is called the radiative zone. As the energy flow gets close to the surface, it becomes more efficient for the energy to reach the cooler surface by causing the plasma to form convection cells. This is similar to the boiling water in a pot, where the water closest to the stove's hot plate is barely moving as the radiant infrared energy streams through it, but closer to the surface the water starts to boil vigorously.

Our sun's radius is about 434,960 miles (700,000 kilometers). The boundary between the radiative and convective zones is 124,274 miles (200,000 kilometers) below the photosphere. This boundary, called the tachocline, is only 18,641 miles (30,000 kilometers) thick but plays a vital role in both generating and controlling the sun's magnetic field. The charged plasma tachocline currents create a magnetic field that is carried to the surface via the solar convection cells, and erupts at the surface as sunspots. These sunspots speckle the surface of the sun in a rising and falling, 11-year sunspot cycle. Other stars with slightly differing convection zones have shorter or longer sunspot cycles that can range from as short as six years for the star Lambda Andromedae to 60 years for the star CC Eridani. Sunspots are dark because the energy that normally would have emerged from this part of the solar surface is diverted away, causing the sunspot plasma to become over 3,600°F (2,000°C) cooler, and so it emits much less light. If you could rip a sunspot from the solar surface and place it in the dark sky, it would have a dull-red glow appropriate to a body heated to 5,432°F (3,000°C).

Above *A coronal loop near a sunspot group revealing magnetic lines of force extending more than 62,137 miles (100,000 kilometers) above the solar surface.*

The surface magnetic field is about ten times stronger than Earth's but in the sunspot cores it can be 10,000 times stronger. The loops of magnetic field can be seen on the sun and resemble the fields around toy bar magnets. Magnetic fields in the presence of plasma are not fixed in space and time, but can release enormous amounts of magnetic energy in the form of flares. These flares can heat the local solar surface to more than 10 million°C and emit X-rays detectable from Earth. Magnetic lines of force can also merge together, a process called magnetic reconnection, which can eject massive plasma clouds into space called coronal mass ejections (CMEs). When this fast-moving plasma cloud reaches Earth, it can cause dramatic events such as the aurora, but also interfere with satellite and electric power technology causing temporary blackouts. Astronomers call these events solar storms, and the many different phenomena triggered by solar events are called space weather.

// How stars are born

Interstellar clouds many dozens of light-years across have densities as low as 10^{-24} g/cm^3, but the centers of most stars exceed 100 g/cm^2. The way in which stars are born is by the gravitational collapse of small volumes of these interstellar clouds. Over the course of a few million years, this collapse causes a 100 trillion trillion times increase in the density of the gas, eventually triggering hydrogen fusion within the compressed and heated core.

The smallest mass in which thermonuclear fusion can take place is about 13 times the mass of the planet Jupiter. Called brown dwarfs, they are able to heat their plasmas so that deuterium nuclei can fuse with hydrogen nuclei (protons) to become an isotope of helium called tritium. At a mass of about 90 times Jupiter, the cores of brown dwarfs are hot enough for hydrogen fusion to start, and the object becomes a true star, which is defined by the mass range from 90 times Jupiter to 100 times our sun, for which hydrogen fusion is the main energy source.

Astronomers have observed many star-forming nurseries in our neighborhood of the Milky Way, and these come in two distinct types. The low-mass star nurseries that produce stars similar to our own sun are dense dark clouds whose interiors include the obscured hot spots of forming stars. These forming "protostars" can be detected with infrared telescopes such as the Spitzer Space Telescope. As the dense interstellar cloud collapses into its individual protostars, the material usually forms a rotating disk of gas. That is because the angular momentum of the gas has to be conserved as it collapses, causing it to spin faster. This is like a spinning ice skater bringing her arms closer to her body causing her to spin faster. The rotating disk contains a magnetic field and so as the gas in the disk falls closer to the central protostar, the field is concentrated and

Below *Size comparisons between the sun, a blue-giant star, and a brown dwarf star.*

Herbig-Haro 24 is located the constellation Orion, about 1,350 light-years away. Two jets of material create clumpy shock fronts that clear out interstellar material. These clumps are called Herbig-Haro objects after the astronomers George Herbig and Guillermo Haro who first discovered them.

Above *At the heart of the nebula NGC 604 are more than 200 hot stars, much more massive than our sun (15 to 60 solar masses).*

amplified. The result is that two magnetized jets of disk matter are ejected along the poles of the rotating protostar to form what astronomers call Herbig-Haro objects. These are seen as luminous clouds of plasma traveling at millions of miles an hour away from the protostar.

Dozens of low-mass stars from about 0.1 to 5 times the mass of our sun are formed in these dark clouds. Their formation is in many cases so gentle that they barely disturb the cloud itself, allowing this low-mass star formation to continue for millions of years. This is an important process for forming what are called open or "galactic" star clusters in which hundreds of stars are formed from the same cloud within a time span of only a few million years. For very massive stars, however, the process is far more violent and terrifyingly beautiful.

Stars more massive than about five times the mass of our sun also form by gravitational collapse, but for reasons not fully understood, this collapse continues until the protostar reaches between five to 100 times the mass of our sun. As the surface of the forming protostar heats up above 10,000°C, the light emitted by the surface becomes richer in ultraviolet radiation until, by a temperature of 30,000°C, nearly two-thirds of all the light is emitted as ultraviolet light. Hydrogen gas can absorb this light but it causes the atoms to lose their sole electron and so become ionized. The amount of ultraviolet light from the star is so intense it can ionize the hydrogen gas within five to ten light-years of the star. The result is a beautiful nebula with multicolor hues representing hydrogen, oxygen, and nitrogen gases being

excited by the ultraviolet light and emitting specific colors. The hydrogen "red" lines and oxygen "green" lines are the most intense colours. Blue colors to these nebulae can occur from the scattering of starlight on dust grains.

Below *The RCW 120 bubble seen by ESA's Herschel space observatory is located about 4,300 light-years away. A star at the center, not visible at these infrared wavelengths, has blown a beautiful bubble around itself with the mighty pressure of the light it radiates.*

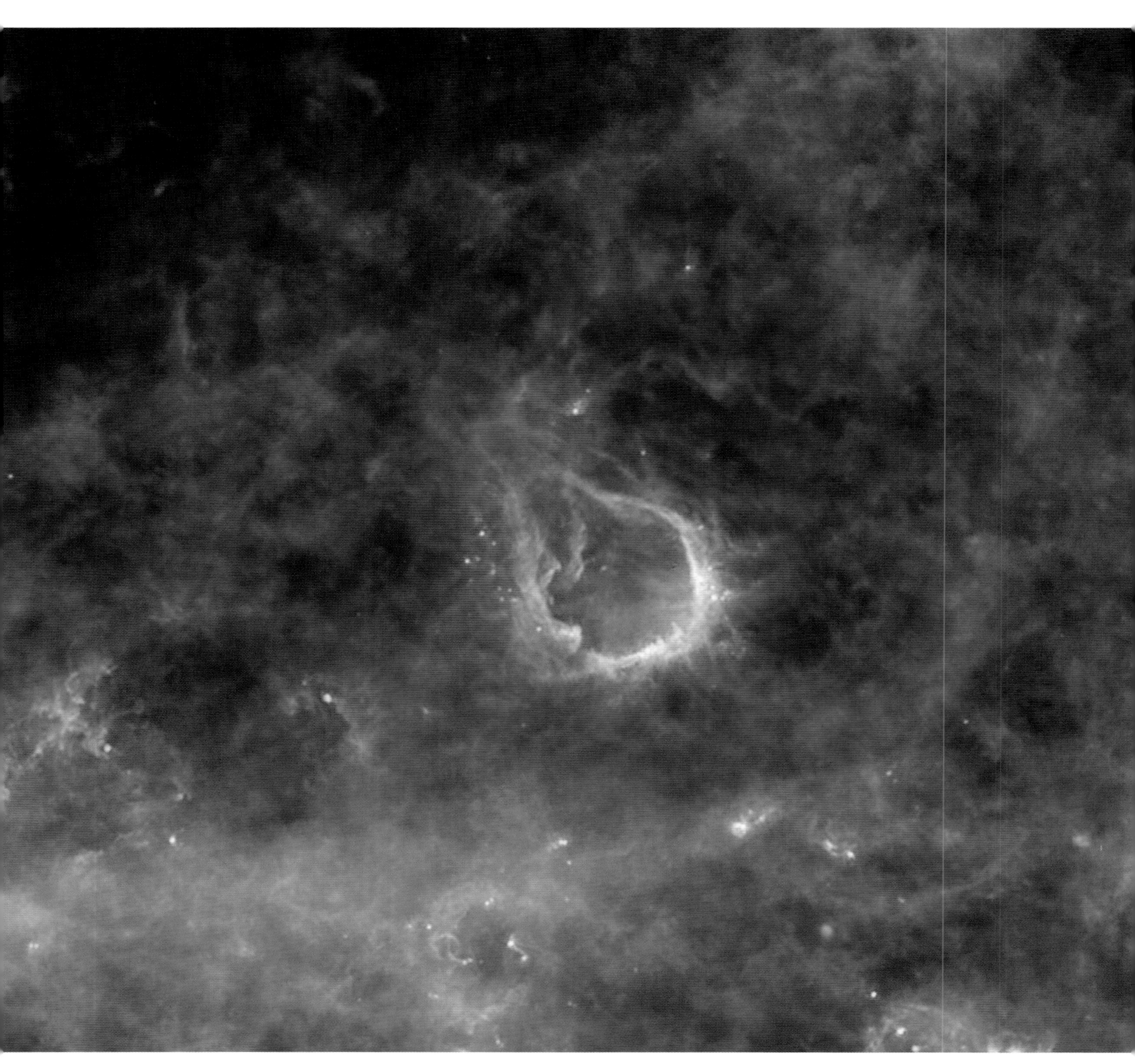

// The middle years

Once a protostar has made the transition to generating its energy by hydrogen fusion, it settles into a long period when it is stable and emitting a more-or-less constant level of light. Over the eons, this level steadily increases as the core of the star burns through its hydrogen reserves. The star's core slowly collapses under its own gravity. This collapse heats the core of the star, but the increased nuclear energy also causes the star to expand, gradually increasing its luminosity. For instance, when our sun first formed it was emitting 25 percent less light and heat than it is now, but has gradually been increasing in luminosity over the last four billion years much like installing a 25-watt bulb in a reading lamp and then coming back in a year to find it has grown into a 40-watt bulb. Currently, the solar luminosity is increasing by 1 percent every 100 million years. An important consequence of this for Earth is that our planet's surface temperature will also increase irrespective of what humans do about global warming. In a few hundred million years there will be no more exposed ice or snow on Earth's surface. By about one billion years from now, the resulting solar and greenhouse heating will cause the oceans to start evaporating. Then, by about two billion years from now, the surface will become uninhabitable. From thermodynamic considerations, there will be no locations on Earth, or under its surface, that will be cooler than the boiling point of water. Multicellular life will become extinct in about 800 million years and the hardiest extremophile bacteria will become extinct in about 1.5 billion years. We are currently about halfway through the era of multicellular life on this planet.

Above *Any starfield will show a variety of stars in different stages of their evolution such as this view of the constellation Orion, where the ancient supergiant star Betelgeuse co-mingles with other stars barely a few million years old.*

The amount of time a star can continue to burn hydrogen to maintain itself against gravity depends on the mass of the star. A star like our sun can burn hydrogen for 12 billion years, but a star 100 times as massive can only do this for a few million. Meanwhile, the far more numerous red dwarf stars with masses of about 10 percent of our sun can last for several trillion years.

During the late-middle years, a star undergoes many

different changes as the quantity of energy produced by the core changes, as well as the location where the energy is being released. For our own sun, in another five billion years the fusion reactions in the core will become less productive as the hydrogen fuel declines, causing the helium-rich core to collapse. At the same time, a hydrogen-rich shell just outside the core will start to reach the temperature where thermonuclear reactions can start. The main source of energy production then leaves the core and resides in what astronomers call the shell-burning zone. This new and more active energy production causes the outer layers of the star to expand and grow cooler, and the star enters its red giant phase. Its "middle-age" years have ended. For stars more massive than our sun, this process is far more complex.

Our sun produces its energy by fusing hydrogen into helium, but more massive stars have hotter core temperatures and so use a more complicated nuclear reaction to produce energy, called the carbon-nitrogen-oxygen cycle (CNO cycle). Instead of the core of the star being only 15 million°C as for our low-mass sun, temperatures in the cores of massive stars can reach 50 million°C. The enormous outpouring of energy allows these stars to be much larger than our sun as they burn hydrogen into helium ash. For example, the nearest O-type star is Zeta Puppis at 1,100 light-years. It has a diameter 18 times our sun and a surface temperature of 40,000°C.

Below *Stars like our sun spend most of their lives as stable stars with slow changes in brightness and temperature, spanning billions of years.*

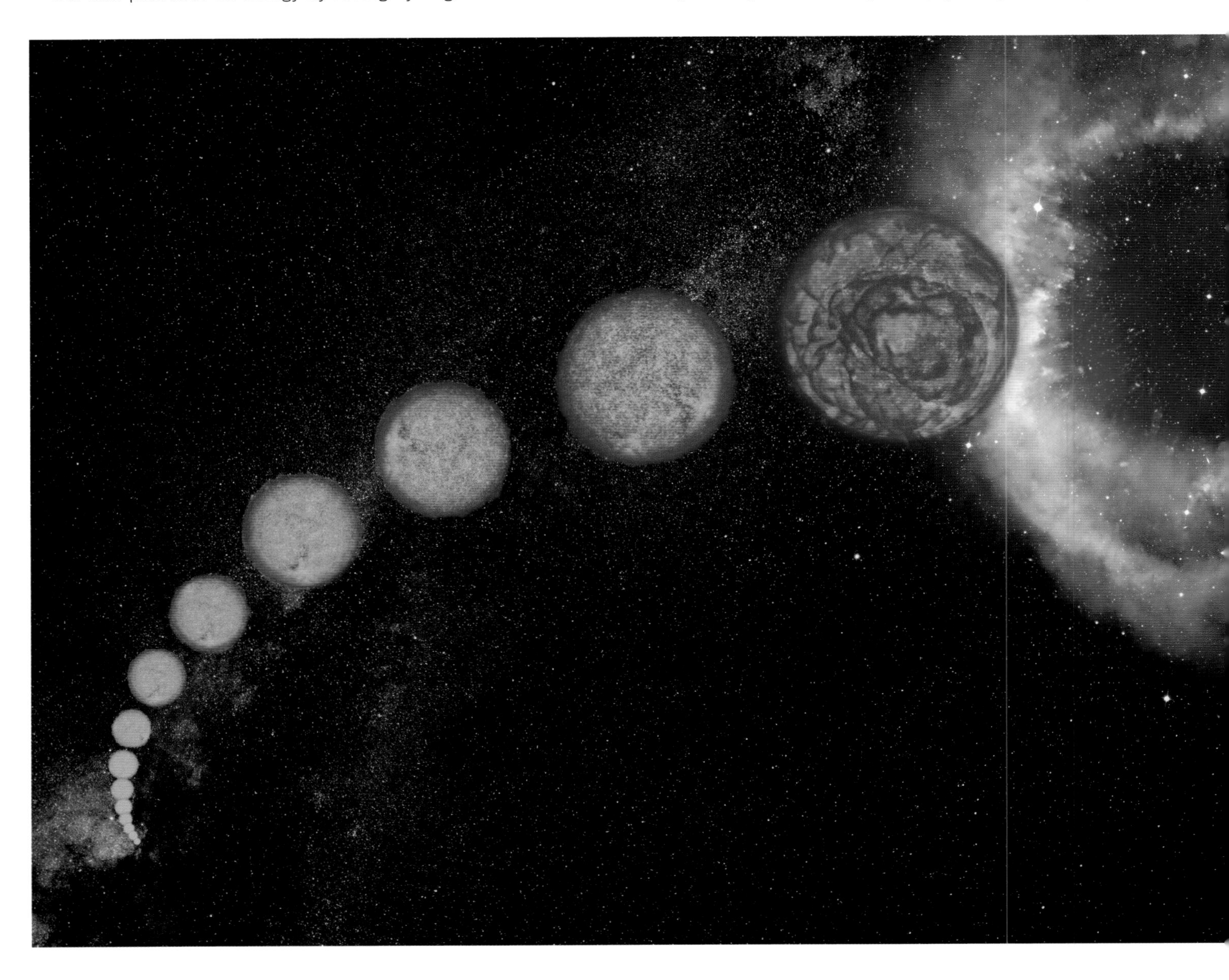

// How stars die

Gravity is relentless, and the available nuclear fuel for supporting the interior of a star is always limited by its mass and its core temperature. Low-mass stars build up a considerable core of helium ash that is inert at the temperatures of hydrogen fusion, however this mass collapses and heats up its surroundings. This process continues until the helium-rich core reaches temperatures near 100 million°C, at which point three helium nuclei fuse into a single carbon nucleus in what is called the triple-alpha reaction. The energy released is dramatically higher than for the hydrogen fusion reaction and the star's outer layers expand and cool so that the star becomes a red giant.

Our sun will enter this stage in about 5 billion years, producing a core of carbon ash inside the fusing helium core left over from the previous hydrogen fusion reaction when it was a middle-aged star. Once the helium ash has been depleted, the remaining carbon ash core will continue its collapse but eventually stop, becoming a solid, Earth-sized object. At this critical point, the outer layers of this low-mass star will continue to expand out into space beyond the orbit of Mars until they escape the sun entirely. For stars less than about four times the mass of our sun, you now have what is observed as a planetary nebula surrounding a brilliant white-hot star at its center called a white dwarf.

White dwarf matter consists of ordinary carbon and oxygen left over as the ash from earlier fusion reactions, and supported by a repulsive force caused by a unique quantum mechanical effect among the electrons. The interior of the white dwarf acts like a gigantic atomic nucleus with electrons, but the electrons have to obey a law in quantum mechanics that says no two of them can be in the same state. This causes what is called electron degeneracy pressure, and this is what holds a white dwarf up against further gravitational collapse. The density of an entire star like our sun crammed into a volume no bigger than Earth is about 1 billion kg/m^3. Its interior is a slowly crystallizing ball of carbon and oxygen at a temperature of 10 million°C. Above its surface, with a thickness of about 6.2 miles (10 kilometers), there is an outer atmosphere of hydrogen and an inner helium layer close to its solid surface. The initial formation surface temperature of a white dwarf is near 100,000°C. Because no further fusion reactions occur, it is steadily cooling over time and will eventually become a black dwarf after hundreds of billions of years.

When a white dwarf is in a binary star system, it

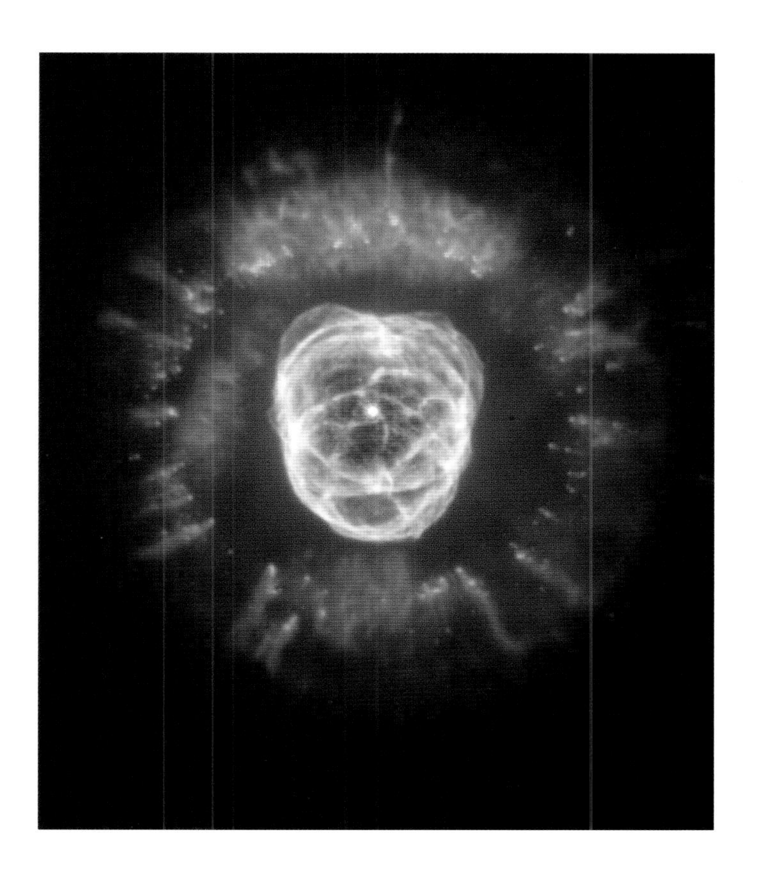

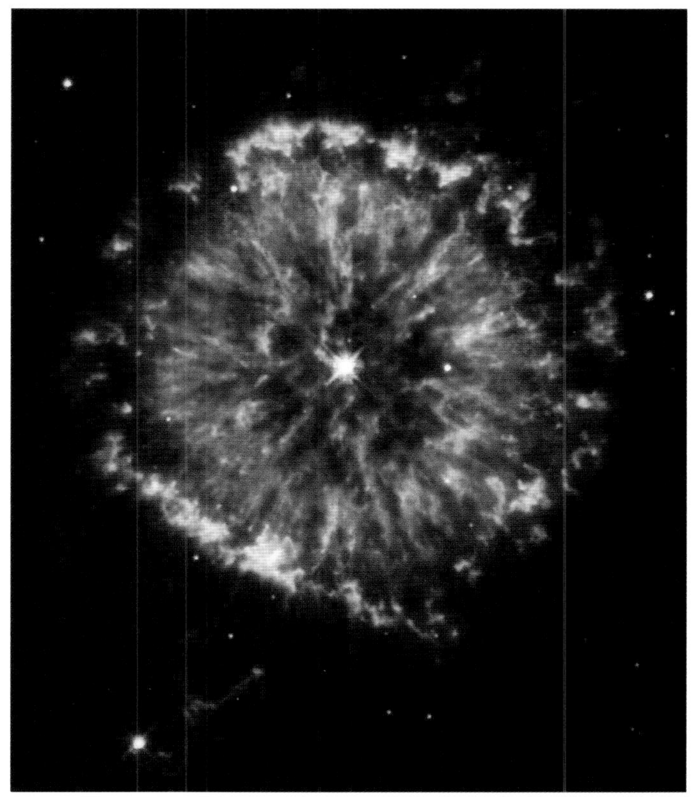

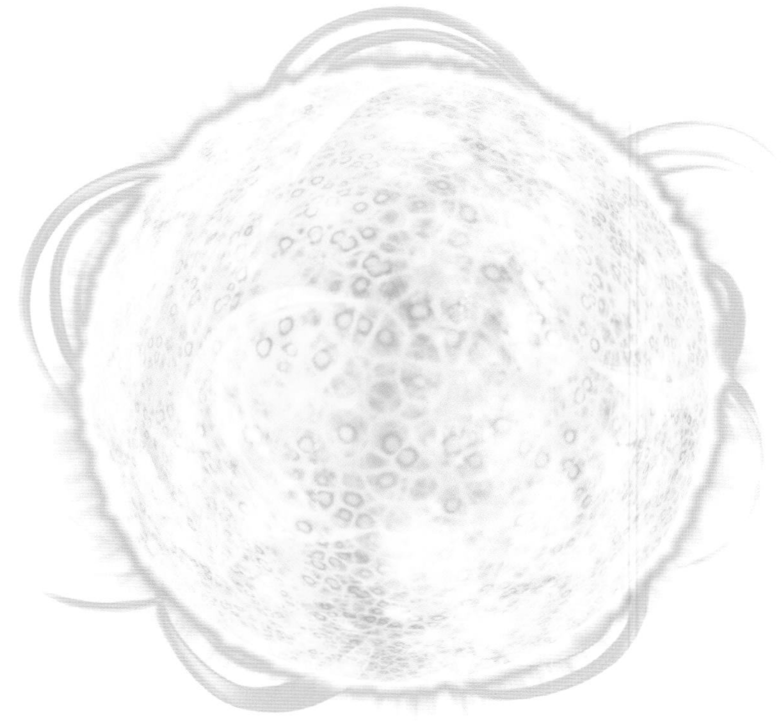

Above *White dwarf stars are about the same size as Earth as this artists' rendering of Sirius B shows.*

Below *The anatomy of a white dwarf showing its composition and shell-like structure.*

Opposite page, top *The planetary nebula NGC 6751, also called the Glowing Eye Nebula, is located 6,500 light-years from the sun. It was formed several thousand years ago and is now about one light-year in diameter.*

Oppostie page, bottom *The double-shelled Planetary Nebula NGC 2392, also called the Lion Nebula, is about one-third of a light-year across and is located 3,000 light-years from the sun.*

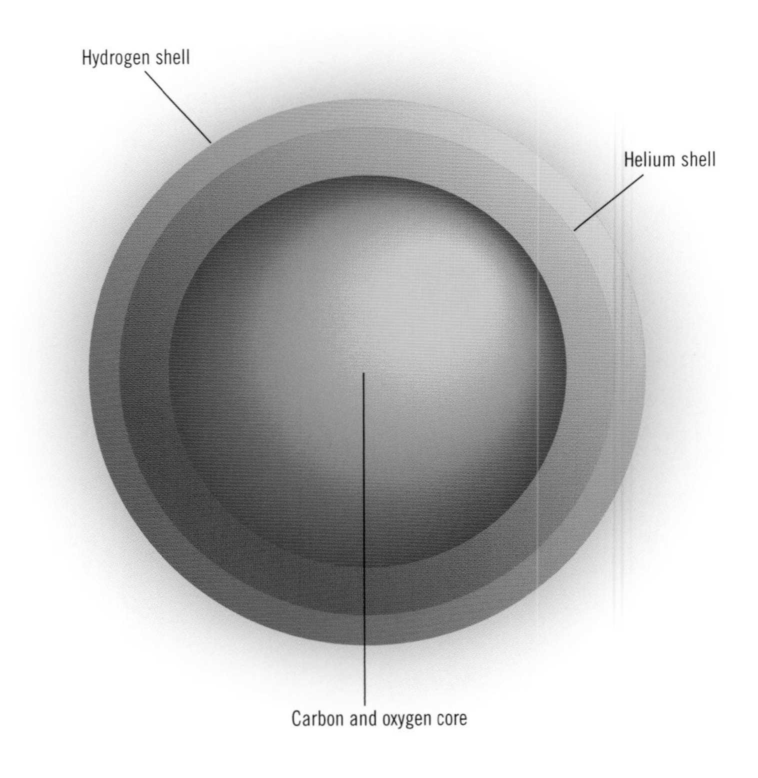

can accumulate additional hydrogen-rich matter from its companion star. When the amount of this matter reaches a critical level, it flashes into thermonuclear fusion causing the white dwarf star to explode as a nova. Under some circumstances, for the most massive white dwarfs and the highest mass transfer rates, this explosion becomes a "Type 1a" supernova and the white dwarf disintegrates completely.

// Supernova

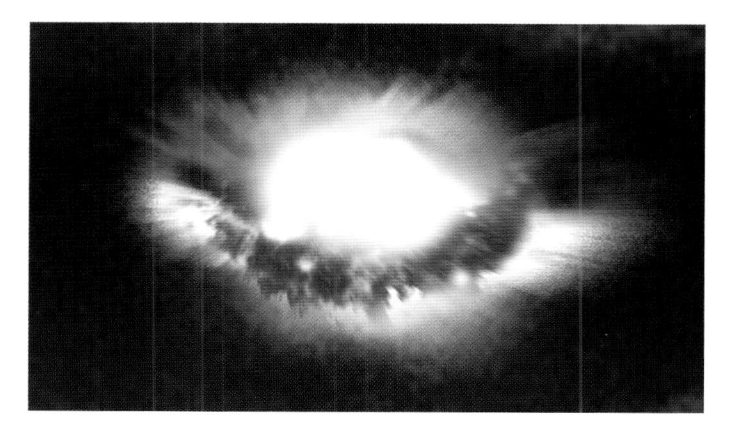

Above and opposite, left to right *An artist rendering of the stages in a supernova explosion involving a white dwarf accreting mass from its companion star and then detonating. These special supernova are called Type 1a and are important "standard candle" objects for determining distances in the universe.*

For high-mass stars, the end of their lives is far more complex and, because of the star's mass, takes much less time. Massive stars can achieve core temperatures far higher than for low-mass stars. When they reach the stage where the triple-alpha reaction has produced a collapsing carbon-rich core of ash, this core can collapse under its own gravity and heat up to temperatures of over 500 million°C. This triggers a fusion reaction called carbon-burning. The energy production takes an order-of-magnitude leap and the star becomes a red supergiant star. The term "burning" is widely used by astronomers to describe the fusion process but in fact has nothing to do with the burning process of chemical reactions. The nearest red supergiant star is Betelgeuse in the constellation Orion only 650 light-years from the sun. The carbon-burning reactions produce such elements as neon, magnesium, oxygen, and sodium plus many other isotopes. In a star like Betelgeuse with a mass of 25 times the mass of our sun, it burns hydrogen into helium for about 10 million years, helium into carbon for about 1 million years, but burns carbon into the other elements for only 1,000 years. Also, although the main loss of energy for low-mass stars is in the light radiation photons streaming out of the core, for massive stars, the largest loss of energy from the core is in the huge luminosity of neutrinos. The neutrinos do not interact with the plasma in the star and so none of their energy goes into supporting either the core of the star or its surrounding layers. This has enormous consequences for what happens next.

At temperatures of 1 billion°C, the oxygen ash burns to silicon, and at 3 billion°C the silicon ash burns to iron. Because iron nuclei cannot be burned to get more energy, this ash builds up as an inert, collapsing core. The density of this billion-degree volume has now reached the point where even the neutrinos have a hard time getting out and so the cooling effect they once had by carrying energy out of the star is suddenly replaced by a heating effect, causing the layers outside the iron core to expand rapidly as a supernova explosion.

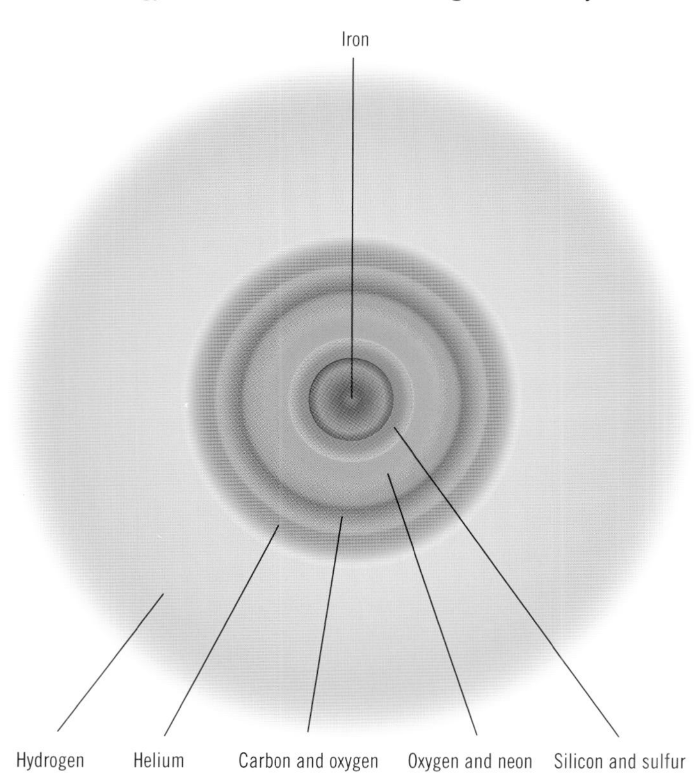

Left *The onion shell structure of a pre-supernova star. As a very massive star nears the end of its evolution, heavy elements produced by nuclear fusion inside the star are concentrated toward the center of the star in a series of layers determined by the fusion reactions and their ashes.*

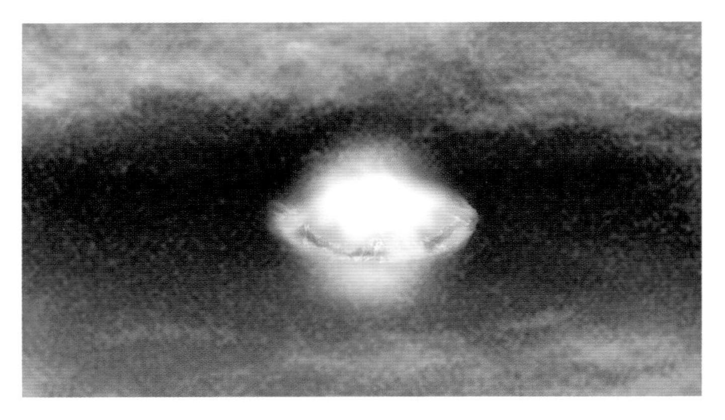

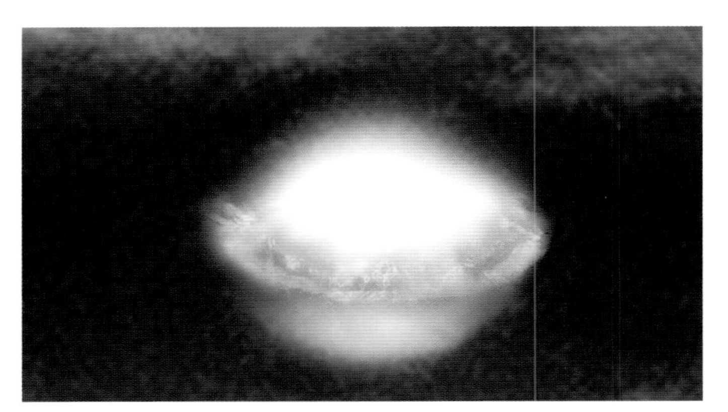

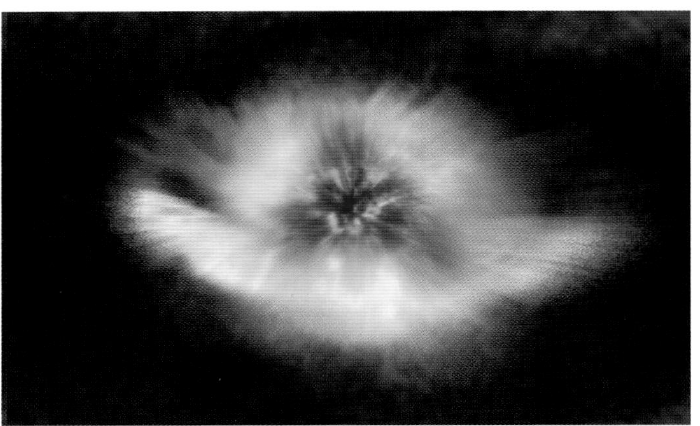

Below *Supercomputer model of an exploding supernova moments after detonation to show the complex motions of matter surrounding the forming neutron star.*

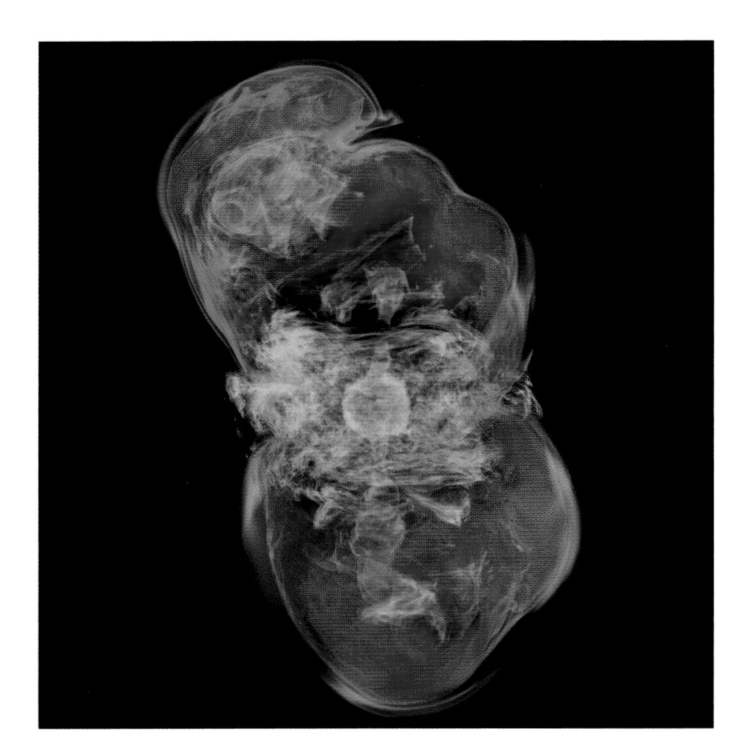

Below *About 600 seconds after detonation, this image from a supercomputer model reveals the formation of a dense magnetic neutron star core called a magnetar, and an outflowing turbulent plasma that ploughs into previously ejected gas to form shock fronts and complex turbulence. The scale of the image is 1.24 million miles (2 million kilometers). Courtesy of Ken Chen, National Astronomical Observatory of Japan.*

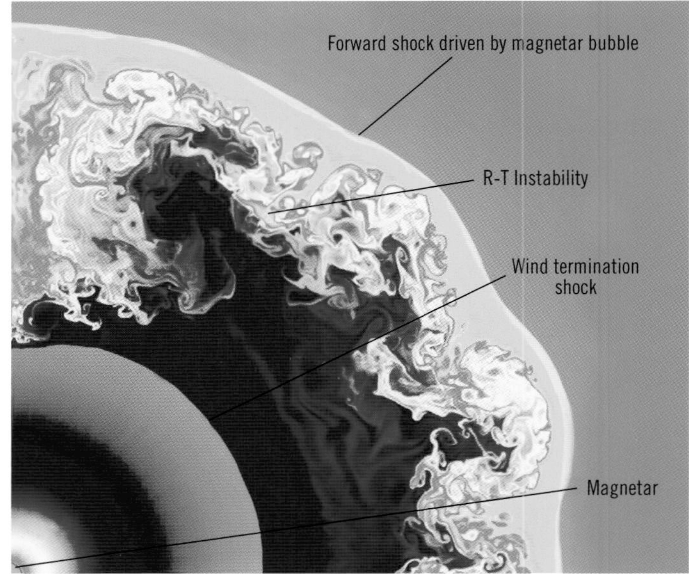

// Supernova remnants

The Crab Nebula in the constellation Taurus is a six-light-year-wide expanding remnant of a star's supernova explosion event that took place in 1054. It was observed by people living in Europe and China.

The explosion outwards from the core produces an equal-and-opposite rocket-effect compression of the iron core to higher densities, leaving behind an object called a neutron star. This will happen if the progenitor star is more than about eight times the mass of our sun. Because the progenitor star was rotating, and like a spinning ice skater pulling in her arms, this rotation speed will dramatically increase as the newly-formed neutron star forms only 12 miles (20 kilometers) across, which can rotate up to 500 times per second. Its surface will also contain a trapped and amplified magnetic field left over from the star that can be trillions of times stronger than Earth's magnetic field. There are thought to be more than one billion neutron stars in our Milky Way galaxy.

A neutron star is the core of a supernova progenitor star that contains about one to three times the mass of our sun but in a volume of space only about 12 miles (20 kilometers) in diameter. These densities are almost exactly equal to that of an atomic nucleus: $4x10^{17}$ kg/m^3. The electron degeneracy pressure that held up a white dwarf is completely overcome by the gravitational collapse and rocket-effect of the supernova detonation. The protons making up the core matter interact with the free electrons in the dense plasma and are converted into neutrons so that, once formed, the remnant dense core is almost 95 percent neutrons, hence the name neutron star.

The interior of a neutron star is extremely exotic. The electron degeneracy pressure that held up white dwarfs is replaced by neutron degeneracy pressure but at a much

Above *An artist's impression of a neutron star accreting gas in a binary system.*

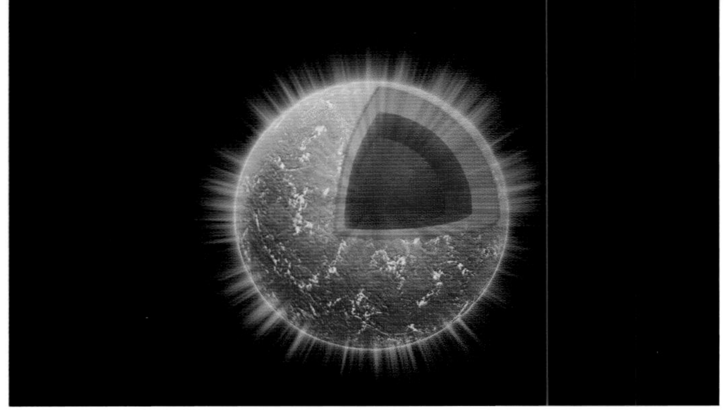

Above *An artist's impression of the neutron star at the center of the Cassiopeia A supernova remnant, about 25 miles (40 kilometers) in diameter. The blue rays represent the neutrinos created in the dense, neutron-rich core.*

higher density. This is because the strong nuclear force is 100 times greater than the electromagnetic force responsible for electron degeneracy pressure. A neutron star has an atmosphere of hydrogen and helium nuclei a few feet thick, laced with a powerful magnetic field that can be trillions of times stronger than our sun's. The interior consists of a thin iron-rich crust only a few hundred feet thick. Below this, and extending to the core, is a nearly-pure neutron material that acts like a superfluid even at temperatures of 1 billion°C.

Neutron stars are so small that, even at their enormous million-degree surface temperatures at birth, they emit hardly enough light to be directly seen from Earth. However, their rapid rotation and strong magnetic fields allow them to be detected as powerful radio beacons across the Milky Way. The first of these "pulsars" was discovered in 1967 by the British astronomer Jocelyn Bell Burnell as a mysterious, repeating radio source. Since then more than 2,000 have been discovered. There are believed to be more than 100,000 pulsars in the Milky Way, with a new one formed from supernova explosions about every 70 years.

Neutron stars are also present in binary star systems when two massive stars individually become supernova and leave them behind. Over time, as they lose energy by emitting gravitational radiation, they eventually collide to become spectacular gamma-ray "burst" sources or even a massive black hole. Supercomputer simulations of these merger events during their last milliseconds reveal collision speeds at nearly the speed of light, emitting huge pulses of gravity waves that can be detected by the Laser Interferometer Gravity Observatory (LIGO) and other gravity-wave telescopes on Earth.

Below *Supercomputer modeling of the last 30 milliseconds of neutron stars merging.*

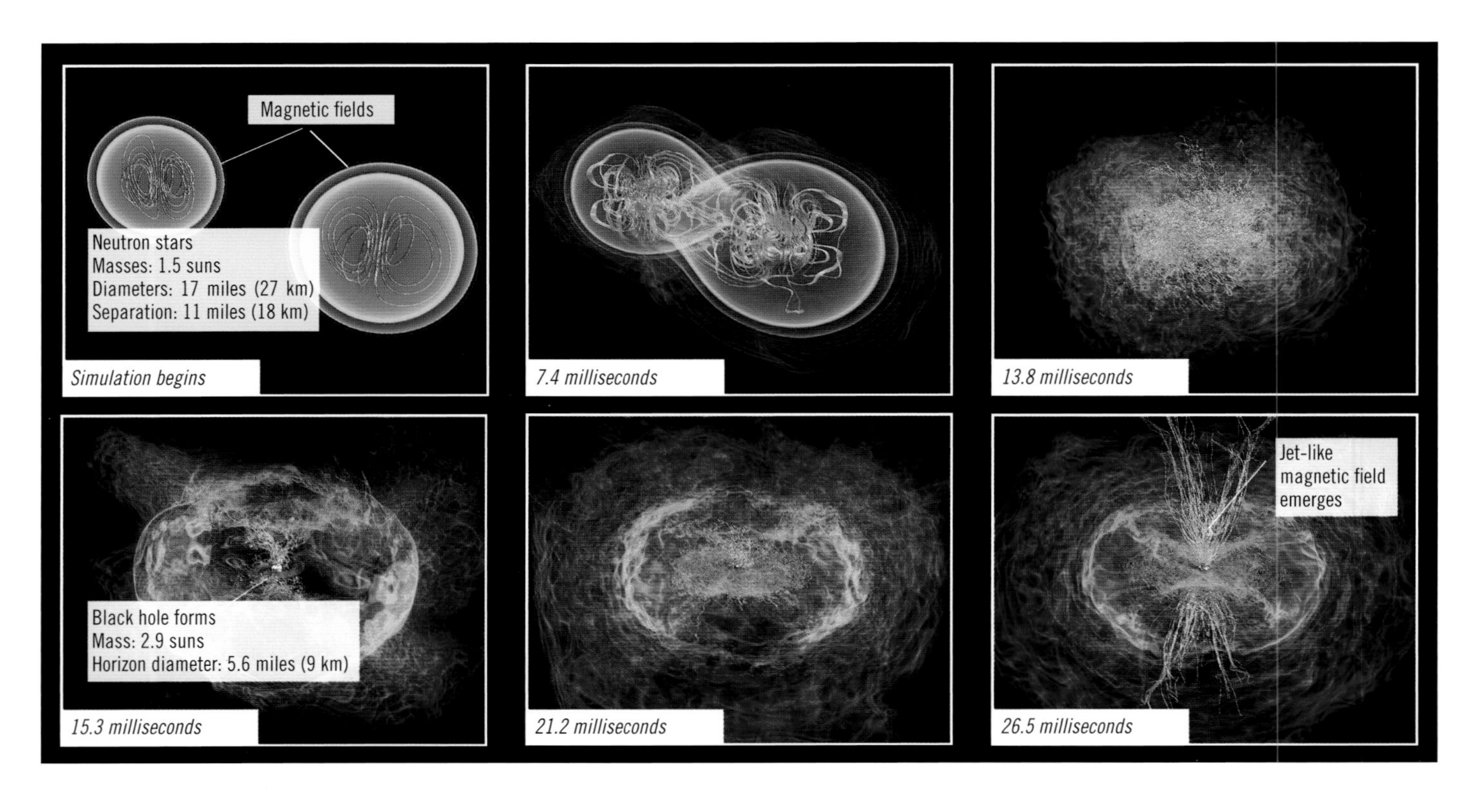

// Black holes

If a progenitor star is more than about 25 times the mass of our sun, its supernova is capable of compressing the remaining core mass well beyond the limits of what neutron star matter can support. Gravity wins the final upper hand and the remnant collapses to become a black hole.

The common definition of a black hole is an object so dense that its own gravity prevents light from escaping from it. Light particles are thought of as miniature rocket ships whose speed is not great enough to exceed the escape velocity of the object, so that light particles fall back and are trapped. This is the kind of interpretation that would work under Sir Isaac Newton's theory of gravity and motion, but it is not correct. In reality, black holes are the product of Einstein's theory of general relativity, which considers gravity to be a feature of the curvature of spacetime. This curvature is determined by an equation that describes how to measure the separations between points in spacetime: a four-dimensional geometry in which time and the three-dimensions of space are co-equal. For a black hole, there is a specific distance where these coordinates take on their usual values outside this distance, but inside this distance the space and time coordinates are actually reversed: the time-like coordinate takes on the properties of a space-like distance, and a space-like distance takes on the properties of a time-like interval. This limiting distance represents the radius of a sphere called the event horizon. Objects falling inside this radius cannot communicate with observers outside the event horizon using light signals traveling at light-speed. Once inside, the object cannot escape because you cannot travel faster than the speed of light. The reason you cannot escape a black hole's event horizon is because spacetime has almost literally put up a wall preventing your escape, or your light signal from getting out. What this means for a supernova explosion is quite dramatic.

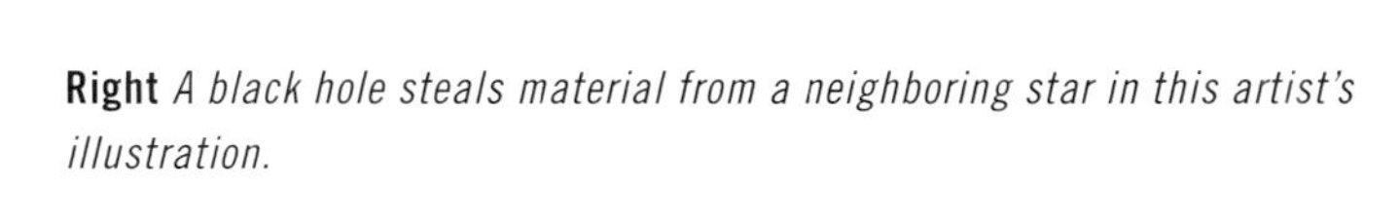

Right *A black hole steals material from a neighboring star in this artist's illustration.*

// Black hole mergers—gravity waves

If you began with a neutron star that is the most massive it can possibly be, it is right at the limit where its neutrons are experiencing their own quantum degeneracy force resisting further compression under gravity. But as you increase its mass still further, its core first dissolves into a gas of free quarks and gluons as the neutrons there dissolve under the enormous pressures. With more mass and external pressure, the speed of these quarks approaches the speed of light and you have now reached the maximum pressure possible from any form of matter. By this time, the size of the neutron star with its maximum possible mass of about two times the mass of our sun is now at a radius of 9.3 miles (15 kilometers). The event horizon radius for that much mass is about 3.7 miles (6 kilometers) if it is not rotating, and is called a Schwarzschild black hole. Adding slightly more mass brings the size of the neutron star equal to its event horizon size, and so the warpage of spacetime literally swallows the object, rendering it invisible except for its gravitational effects.

Contrary to common descriptions, black holes do not act like vacuum cleaners and "suck in" any material in their surroundings, but they do cause nearby matter to lose orbital energy as they emit gravitational waves. This loss of energy causes the orbits to continually creep closer to the event horizon and final one-way absorption by the black hole. Surrounding the black hole event horizon out to a distance of about twice the horizon radius is a zone in which objects there cannot find stable circular orbits to maintain a constant distance from the black hole. Orbit decay takes over and the particle inexorably slides through the event horizon. Even light signals that graze this zone can be captured into temporary circular orbits and circulate for a period of time. Because black holes "warp" spacetime, they also behave like optical lenses and can bend light rays passing by them with many interesting effects. Thanks to the equations of general relativity, astronomers can look at the distorted imagery surrounding a black hole and recover what the background object looked like. This has been a powerful technique that lets astronomers study the distant galaxies in the universe by using massive foreground clusters of galaxies as a gravitational lens.

Right *This artistic interpretation shows two black holes on a collision course, each surrounded by their own swirling accretion disk of matter.*

// The gravity wave universe

The most massive neutron star is about twice the mass of the sun, but despite how common they should be, astronomers have not been able to detect numerous black holes smaller than a mass of about three times the sun's mass. One possibility is these large black holes are not formed by supernova that produce slightly more massive neutron stars, but by some other mechanism. One of these is black hole mergers. There are several possibilities for this: two black holes could merge to form a larger one; a black hole and a neutron star could merge; or a black hole and a white dwarf could merge. It is also possible for two neutron stars to collide and merge. Astronomers have over the decades found many examples of binary neutron star systems such as PSR J0737-3039, or neutron star-white dwarf binary stars such as J0453+1559, so the idea that mergers of these remnants could form the black holes is not that far-fetched. In fact, one of the most amazing confirmations of Einstein's general relativity is that objects produce gravity waves when they are accelerated, and the most severe accelerations should come during these merger events.

Below *Colliding black holes send ripples through spacetime that can be detected on Earth. The Advanced Laser Interferometer Gravitational-Wave Observatory, which has detectors in Louisiana and Washington, has directly observed these gravitational waves.*

Above *The VIRGO observatory in Louisiana is the companion facility to the LIGO facility in Washington. They operate as a pair to detect gravitational wave events.*

Physicists have hunted for signs of gravity waves for decades using a wide array of different technologies and strategies. In 1994, the Laser Interferometer Gravity Observatory (LIGO) began construction in Hanford, Washington, and Livingston, Louisiana, at a cost of $1.1 billion. Following several design and sensitivity changes, on September 14, 2015 it detected the first clear and unmistakable signal of a distortion in the geometry of spacetime. The event lasted only half a second, but, with general relativity, physicists could determine it was produced by two, 30 solar mass black holes that had merged in a distant galaxy some 1.3 billion light-years from the Milky Way. By 2020, LIGO had detected over 20 events, all related to mergers of one kind of compact object with another. One of these is a neutron star–neutron star collision, GW170817, on August 17, 2017 each with 1 to 2.5 solar masses of material.

Now that we have explored stellar evolution and the various key events that mark this process we can resume our story about the evolution of our universe.

THE FIRST STARS AND GALAXIES

We know that the stars around us formed from collapsing clouds of gas, but how did the earliest stars in the universe form? Astronomers are searching for the youngest stars and galaxies that formed after the Big Bang to learn about this formation process, and whether the stars were massive enough to create the heavier elements beyond beryllium such as carbon, oxygen, and silicon. Without these elements, life would not be possible in this universe. Astronomers predict that, with only hydrogen and helium, these earliest stars were over 100 times the mass of our sun and exploded as supernovae within a few million years. Their remains, strewn across the infant universe, can be found in the carbon, oxygen, and iron that make up our bodies today.

An artist's impression of the galaxy A2744_YD4 from when the Universe was just 4 percent of its current age.

// Population III—the first stars

We think that the first stars began to form in the early universe much the same way they do today. One of the biggest differences is that these stars formed from cool gas that was essentially pure hydrogen and helium. Today's stars by comparison are made from hydrogen and helium, but they also contain up to 2 percent by mass of the other heavier elements as well. Astronomers call this the "metallicity" of the star, because the metal iron is the most common element to be detected in stars and is used as a gauge to classify them. Our sun has a metallicity of 0,012 or 1.2 percent meaning that 1.2 percent of its mass is in elements heavier than helium. Metallicity has the important effect of making the interior of a star less transparent to light. The opaqueness of the matter causes radiation pressure to stop the in-fall of matter at a lower mass as you increase the metallicity. When the star was forming from in-falling material, this growth process led to high-metallicity stars with smaller masses. For the first generations of stars made from pure hydrogen and helium, their metallicities were essentially zero. This means that the typical stars that formed had very large masses, perhaps up to 500 times the mass of our sun. By comparison, the most massive stars in our Milky Way today are R136a1 (315 suns) and R136c (230 suns) located in the Tarantula Nebula 163,000 light-years from Earth.

A second consequence of the low metallicity is that these massive stars produced enormous amounts of thermonuclear energy in their cores, and their surfaces were over 100,000°C hot. Objects this hot produce more than two-thirds of their light in the ultraviolet part of the electromagnetic spectrum. There are stars in our galaxy that behave this way, such as the stars in the heart of the Orion Nebula. One of the biggest known consequences of this UV light is that it will ionize hydrogen atoms for many light-years in all directions.

The light from these ancient and massive stars, called Population III stars, is invisible from Earth but should be visible at infrared wavelengths by new generations of telescopes such as the Webb Space Telescope. Its infrared sensors are specifically designed to detect the tell-tale infrared light from these massive stars still buried inside the primordial gas clouds from which they were formed.

Right *The Tarantula Nebula is about 1,000 light-years across and is ionized by a cluster of massive stars at its core in the asterism known as R136. With a mass of over 400,000 times our sun in gas, it also contains 72 massive stars in a cluster only 20 light-years across. Only two million years old, it contains nine stars, each with more mass than 100 suns. Trillions of these objects filled the universe when the first generations of stars were being born.*

// Ultraviolet light floods the universe

Astronomers expect that there weren't just a few of these massive Population III stars sprinkled across the universe at that time, but uncountable trillions of them. The effect they had on the dark era was nothing less than phenomenal. Within a few tens of millions of years, the ultraviolet light pouring out of these stars flooded just about every corner of the universe and, over time, re-ionized all of the hydrogen and helium gas that was present in space. The dark era ended and was replaced by a mottled patina of massive nebulae resembling the Orion Nebula but at vastly larger scales than common nebulae in our galaxy. Astronomers call this the re-ionization era because, for the second time in the history of the universe, the primordial hydrogen and helium gas was ionized into a plasma of nuclei, protons, and electrons. This event may have started about 250 million years ABB and may not have been completed for another 500 million years.

During the re-ionization era, the formation of massive stars was becoming more and more intense as time went on. The galaxy SSA22-HCM1 detected at an age of one billion years ABB was producing new stars at a rate of 40 solar masses per year. Within a few hundred million years, only small galaxy-sized blobs of gas still remained in the now fully-ionized universe and were being quickly evaporated. We can still see the ghosts of these clouds in the light from very distant galaxies, which astronomers call Lyman-alpha blobs or LABs.

LABs are huge concentrations of a gas and are some of the largest known individual objects in the universe that have been detected from that time. Some of these are more than 400,000 light-years across. The Himiko LAB is 55,000 light-years across, which is half of the diameter of the Milky Way. We are seeing its light when the universe was only about 830 million years old.

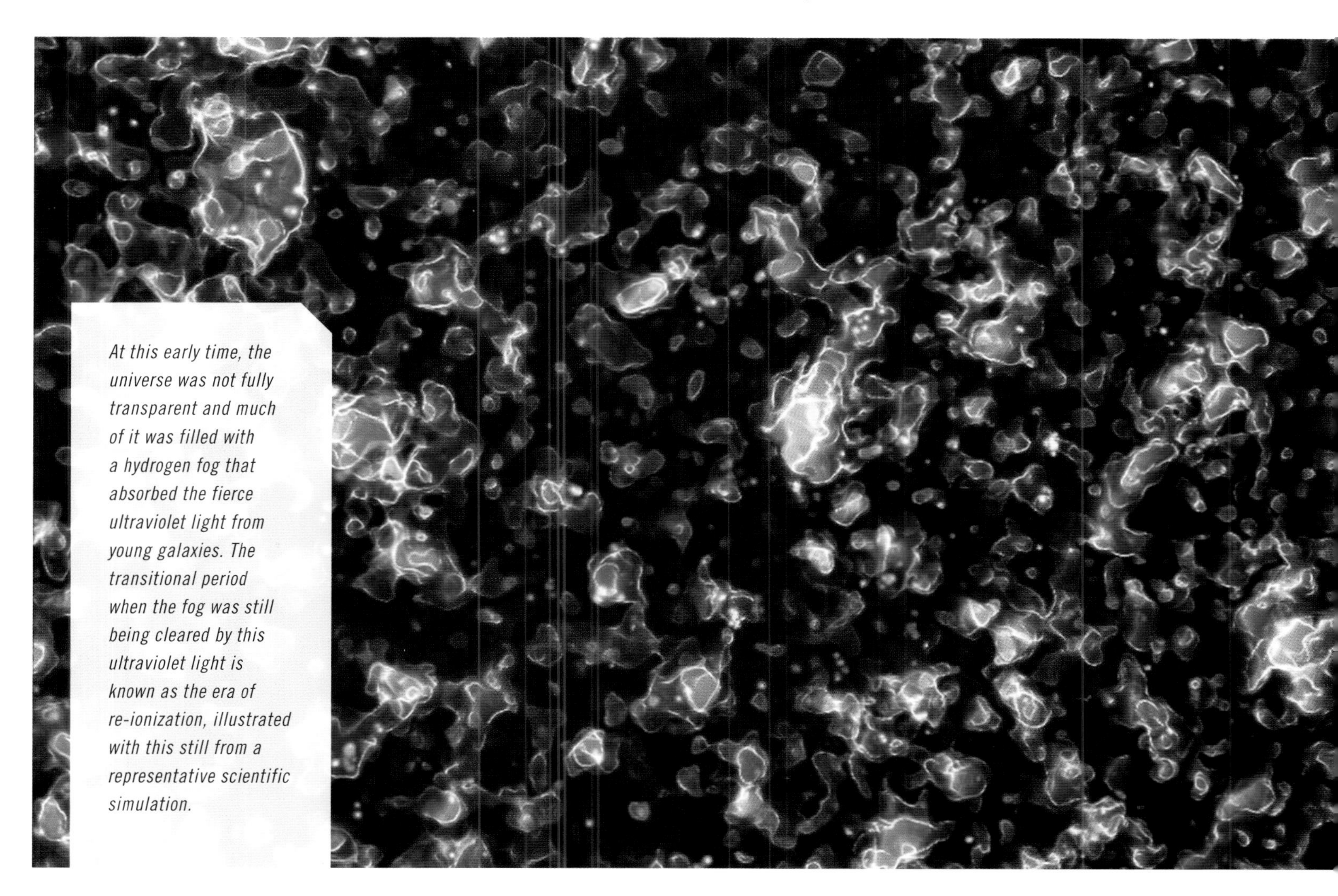

At this early time, the universe was not fully transparent and much of it was filled with a hydrogen fog that absorbed the fierce ultraviolet light from young galaxies. The transitional period when the fog was still being cleared by this ultraviolet light is known as the era of re-ionization, illustrated with this still from a representative scientific simulation.

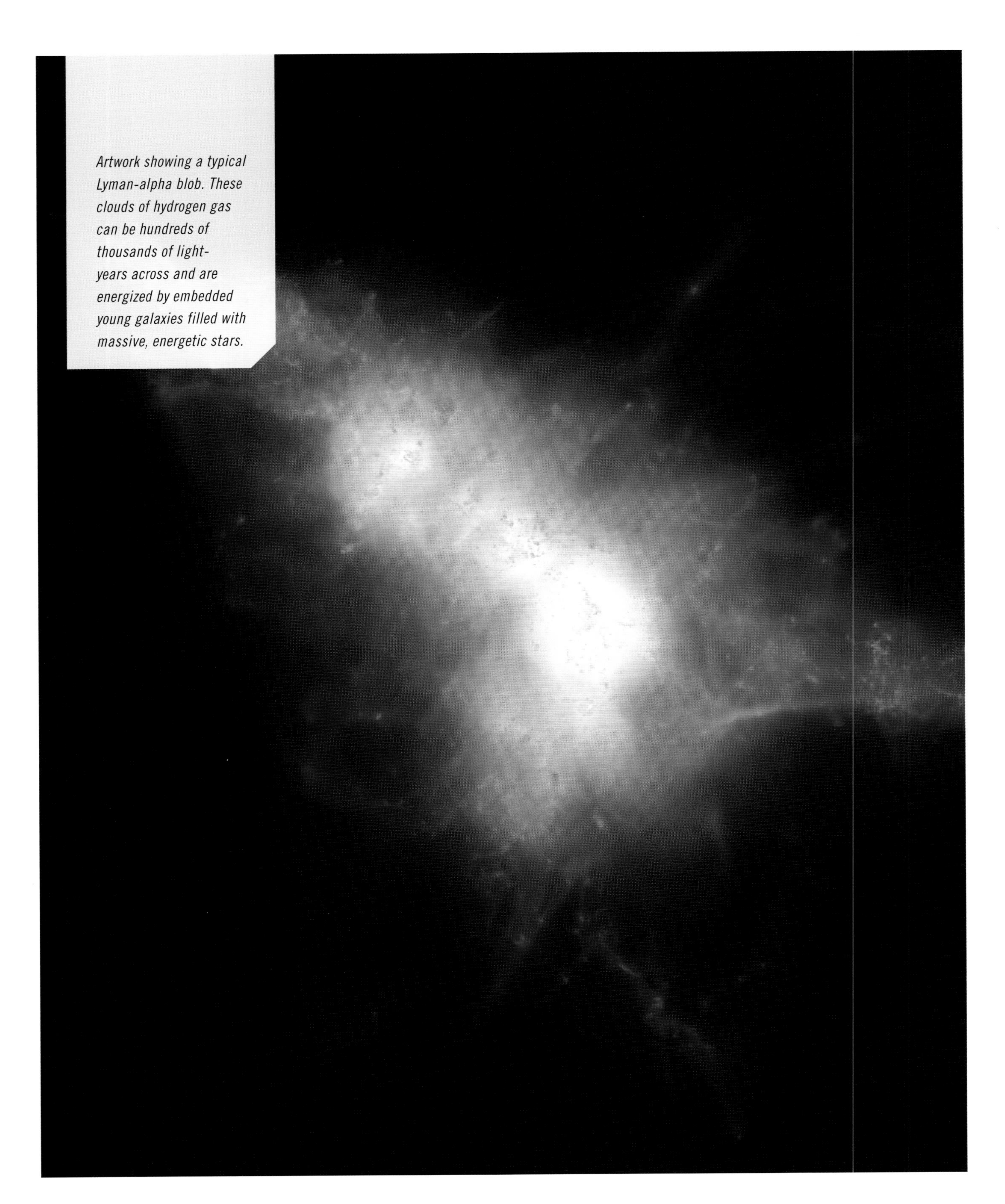

Artwork showing a typical Lyman-alpha blob. These clouds of hydrogen gas can be hundreds of thousands of light-years across and are energized by embedded young galaxies filled with massive, energetic stars.

// Element enrichment supernovae

The third consequence of having stars this massive is that their lifespans from formation to supernova is measured only in millions of years. When they reach the ends of their lives, they become supernovae, but for stars more than 100 times the mass of our sun, astronomers have coined a new name for their detonations: hypernovae. The remnants of these titanic hypernova explosions is the formation of black holes and equally deadly gamma-ray bursts that can beam enormous amounts of light energy up to billions of light-years from the star. They also spew out into space the processed nuclear matter from their deep interiors. This matter has been enriched with elements heavier than helium such as carbon, oxygen, and iron that did not exist in the universe thus far.

This process of element enrichment can be observed today in the remains of many supernovae that have been studied in detail. Astronomers can use sensitive spectroscopic techniques to identify the elements in any glowing collection of interstellar or intergalactic gas and determine element abundances directly. For example, as viewed from Earth, the supernova called Cassiopeia A detonated sometime in the 19th century. Spectroscopic studies later discovered that it had ejected 10,000 Earth masses of sulfur and 70,000 Earth masses of iron. In fact, essentially all of the elements needed to create planets and DNA molecules have been detected in the remains from this one typical supernova. With trillions of Population III supernovae in the early universe, the primordial gases could have been quickly enriched to varying degrees with the heavy elements in only a few hundred million years.

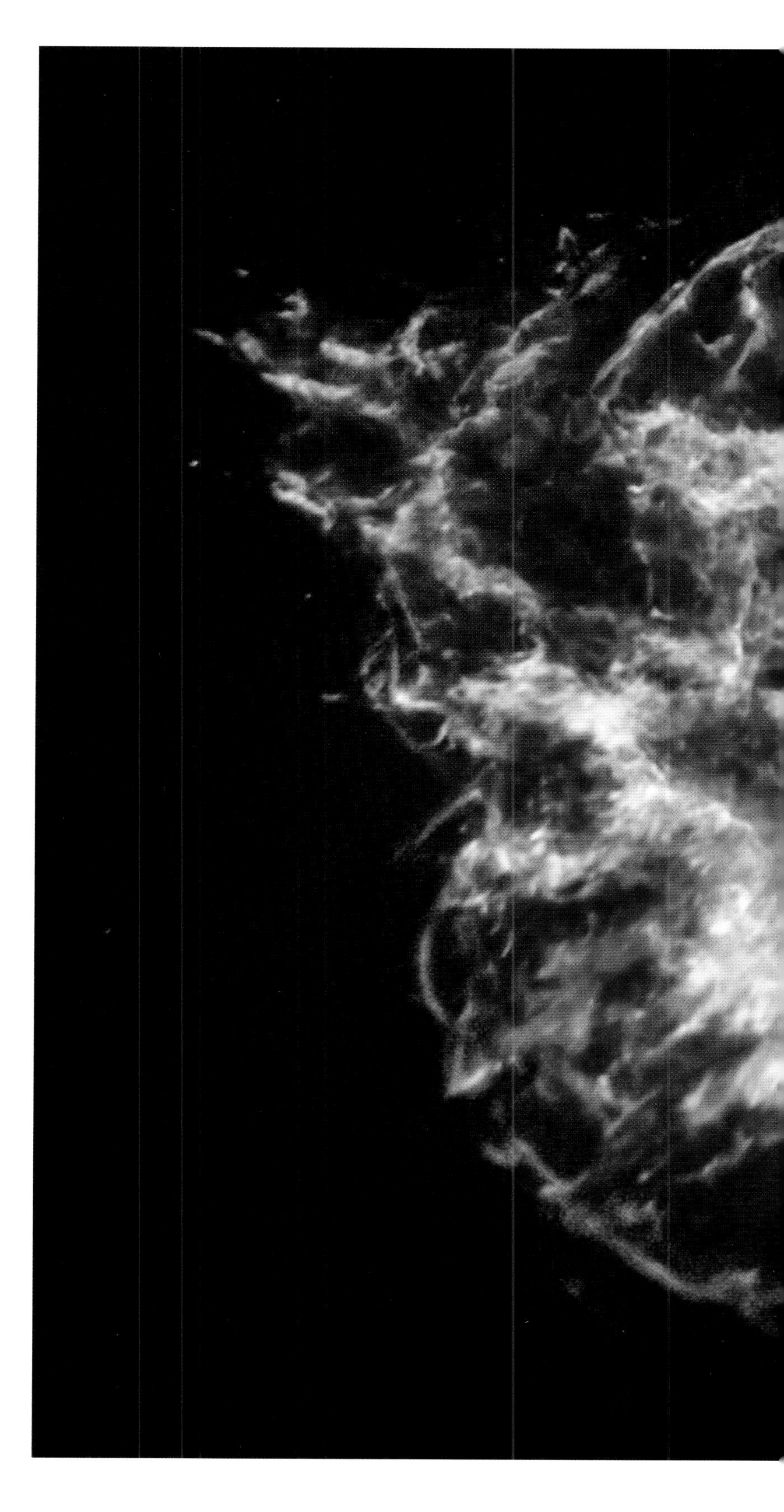

Right *A Chandra Observatory composite image of the supernova remnant Cassiopeia A, showing light emitted by atoms of silicon (red) and iron (blue) in the shocked outflowing gas cloud.*

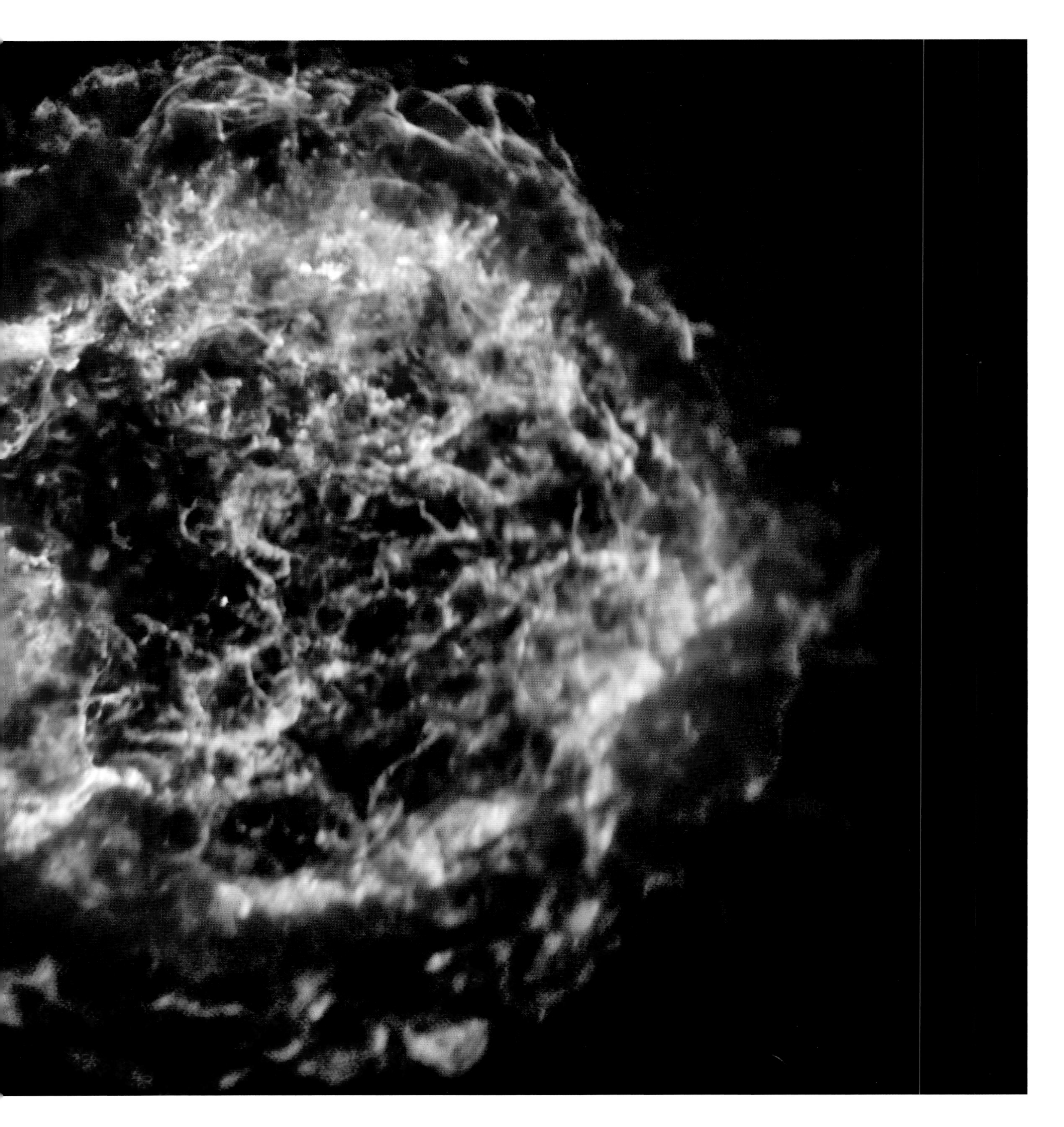

// Young galaxies

Astronomers believe that, like the massive stars in our Milky Way today, the massive Population III stars in the early universe tended to be formed together into clusters of stars. The interaction between these clusters and the surrounding gases was not merely to ionize them, but to aid in the formation of other stars. Over time, small galaxies began to take rudimentary shape as collections of millions of solar masses of gas spawned thousands of star births per year. Today these collections resemble dwarf galaxies like the companions to our Milky Way, the Magellanic Clouds. It is very difficult to see these young galaxies with telescopes even as powerful as the Hubble Space Telescope. They appear as tiny smudges of light with spectra that suggest enormous distances, and so are seen as they were when the universe was only a few hundred million years old.

NASA's Webb Space Telescope, launched in October 2021, lets us see these distant and intriguing blobs of light

Below: *An artist's rendering of the Population III stars detected in the young galaxy CR7. These primordial stars were born from clouds of gas that contained hydrogen and helium (and trace amounts of lithium), but no heavier elements.*

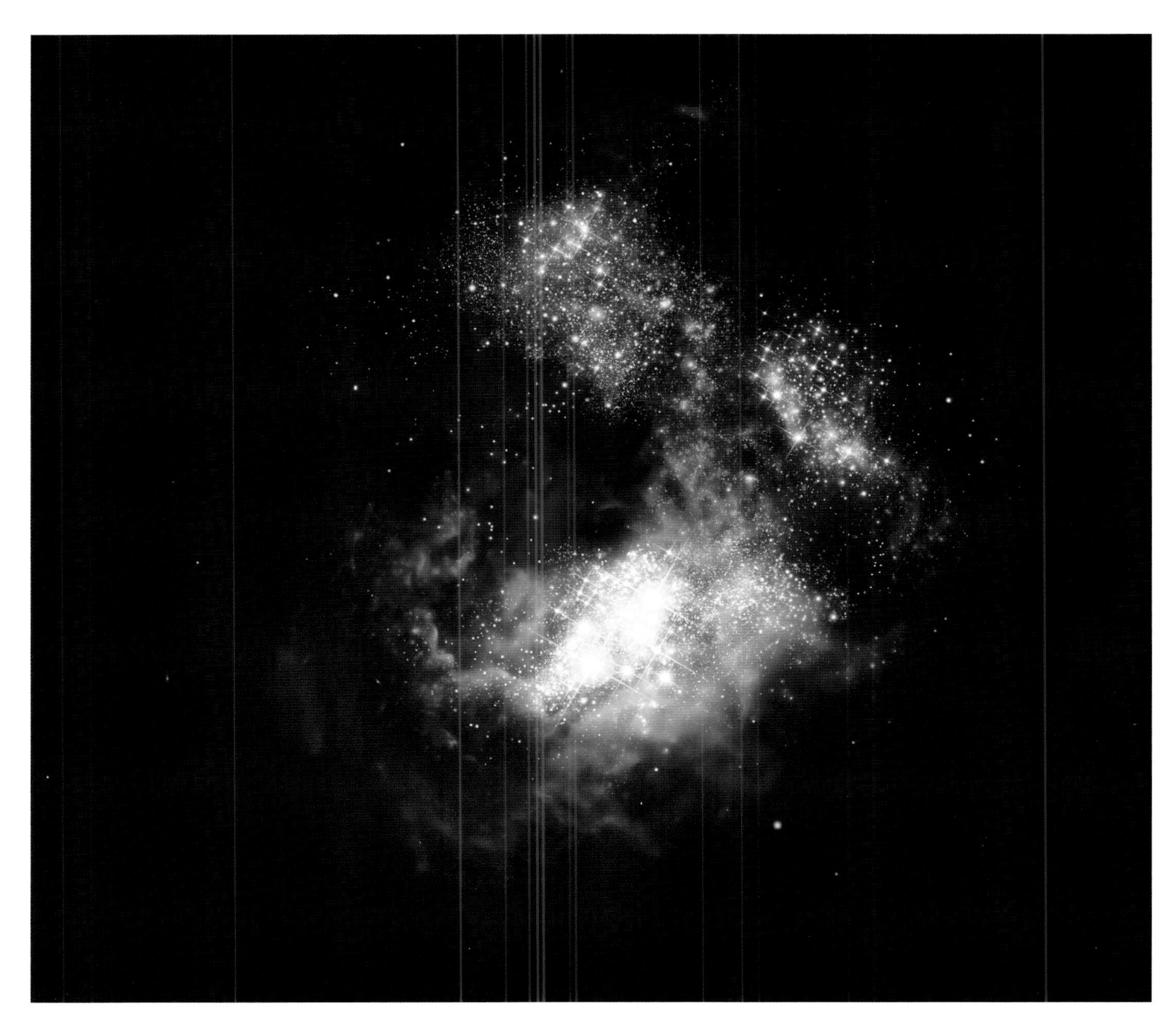

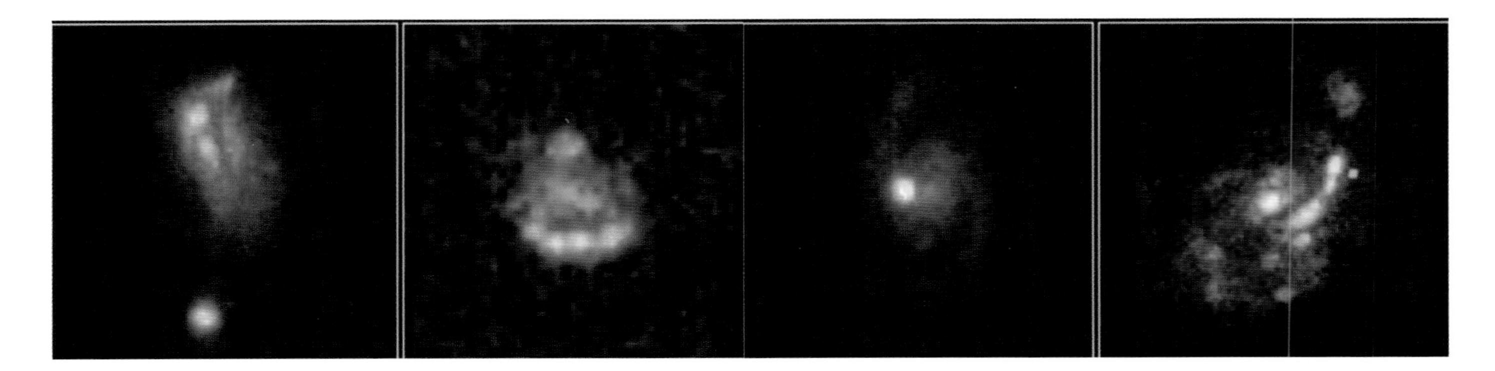

Above *A selection of normal spiral galaxies estimated to be at a distance of 5 to 7 billion light-years from the sun. These galaxies existed when the universe was half its present age. Their irregular and clumpy shapes indicate galaxy collisions and mergers actively taking place.*

Below *Markarian 209 is a dwarf galaxy in the constellation Canes Venatici. Within it, two giant star clusters appear brilliant white and are enveloped by greenish hydrogen gas clouds. This dwarf galaxy is 15,000 light-years across and contains about one million suns'-worth of gas and stars.*

in far more detail, but we may not have to wait for very long. By comparing the properties of these infant galaxies with nearby galaxies we can get a good idea of what they might look like close-up.

Astronomers have been studying galaxies called Lyman Break Galaxy Analogs (LBGA), which are distant enough to resemble the UV spectrum, size, and star formation rates of the far more distant infant galaxies, but are only about 1–2 billion light-years distant and so can be seen in some detail. They resemble spiral or irregular galaxies but with a very clumpy appearance and often seem to be interacting with other nearby galaxies. It is likely that the very young galaxies looked similar to these LBGAs but with much larger numbers. Perhaps all present-day galaxies passed through this lumpy stage of star-forming activity before they settled down to the systems we see around us today.

// Supermassive black holes appear and grow

As the earliest generations of stars exploded in hypernovae and produced black holes, there was plenty of opportunity for these black holes to collide and grow into larger black holes containing hundreds or thousands of times the mass of our sun. But all things take time, and what is very surprising is how quickly this population of black holes grew to gargantuan sizes in seemingly too-short a time, even by successive mergers.

How do we discover these supermassive black holes? Astronomers do this by looking for signs of young objects in the ancient universe that are far brighter than can be understood simply from the light of the Population III stars alone. In the current universe, these objects are called quasars, because their nuclei are producing thousands of times more light energy than entire galaxies like our Milky Way. The prevailing idea is that each quasar core is itself a supermassive black hole with more than one billion times the mass of our sun that is at the center of a rotating disk of gas. The gas and other debris in this disk, which can be more than 300 light-years across, is flowing into the black hole and releasing huge amounts of energy. So, astronomers look for the most distant objects they can find, which are the youngest and most luminous objects in the universe, and see if they behave like quasars.

The most distant known quasar, discovered in 1998, is called APM 8279+5255. We are seeing its light from a time when the universe was about 1.6 billion years old. It is producing about 100,000 times the light from our own Milky Way. The mass of this supermassive black hole is about 23 billion times more massive than the sun. Astronomers do not know how a black hole this massive could have formed so soon after the Big Bang. This would imply a phenomenal black hole growth rate higher than 15 solar masses a year! A still-larger ultra-massive black hole is called TON618. At a distance of 10 billion light-years, it produces more light than 140 trillion suns, and has a mass of 66 billion suns. This means that over its 3 billion-year lifespan, it has been accumulating mass at an average rate of 22 solar masses each year.

Right *The ultra-massive black hole Holmberg 15A is 700 million light-years from Earth but has grown to over 40 billion times the mass of our sun since the Big Bang. Its X-ray light can be detected from Earth by space observatories such as NASA's Chandra.*

Above *The star associated with the Cygnus X-1 black hole, obtained by the NASA Chandra Observatory. The companion black hole is invisible but orbits this blue supergiant star, which has a mass of 20 times our sun.*

IMAGING A SUPERMASSIVE BLACK HOLE

Black holes were first proposed by astronomers in the 1960s as a new way of producing huge amounts of energy in a very small volume of space. The challenge is that they can only be detected by their gravitational influences on nearby stars and gas. Star-sized black holes have been discovered in orbit around a number of different stars; the closest of these is Cygnus X-1 at a distance of 3,000 light-years. Another well-studied black hole is at the center of our Milky Way some 26,000 light-years away. In neither case can the black hole itself be directly seen. Instead, you study the gravitational pulls they exert on nearby stars. But in 2019 that changed.

Astronomers using eight radio telescopes across the world were able to combine the data and use it to form the first image of a supermassive black hole. It was located in the nearby galaxy Messier 87 in the constellation Virgo about 53 million light-years away. The existence of this black hole, with a mass of 6.5 billion times our sun, had been suspected for decades by a variety of independent studies of this powerful radio galaxy. The image formed from trillions of bytes of data over the course of several months of observation revealed many essential details of the accretion disk and the black hole.

As expected, it is surrounded by a luminous disk of gas that is orbiting the supermassive black hole at over 1,000 km/sec or 2 million mph. The image looks like a luminous donut in space because the gravity of the black hole is so strong, it is bending the light from the disk behind the black hole so that it forms a round circle from our perspective. The diameter of the ring is about 0.01 light-years or 36 times the diameter of our solar system out to the orbit of Pluto. The dark hole at the center contains the black hole itself, which has a mass of about 6.5 billion times our sun. This means that the radius of its event horizon is 12 billion miles (19.5 billion kilometers) or about three times the radius of our solar system. A detailed study of the geometry of the ring also shows that this supermassive black hole is spinning, and from our perspective rotating clockwise. This is only the first of several hundred other supermassive black holes that astronomers will directly image in the coming decades using the Event Horizon Telescope. Astronomers also hope to take time-lapse movies of matter circulating in the surrounding disks and perhaps even flowing onto the black holes themselves.

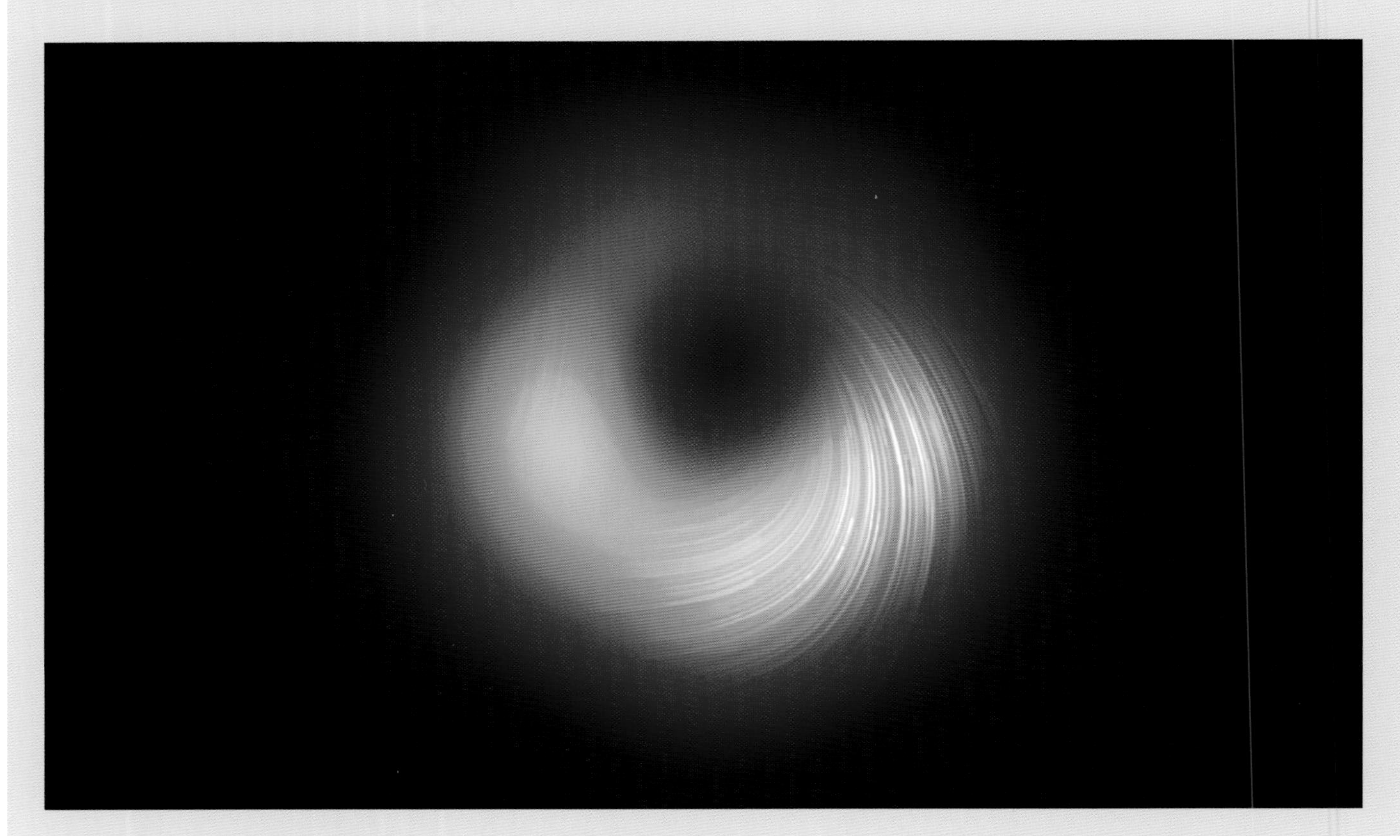

Above *The supermassive black hole in the active galaxy Messier 87 was imaged by the Event Horizon Telescope radio observatory consortium and reveals the spiral pattern of magnetic fields embedded in the inflowing gas.*

// Galaxy groups and clusters start to become common

Following the hidden gravitational wells implanted on the cosmos by dark matter, ordinary matter had no gravitational choice but to follow along. As stars and galaxies began to take form out of the smaller gravitational tide pools, they followed the larger-scale patterns of dark matter to assemble into clusters of galaxies and even larger filamentary shapes and voids. We can still see this imprint of dark matter by mapping out where all of the galaxies are in our corner of the universe. Once the dark age and the re-ionization era came to an end about one billion years ABB, all that was left were the luminous bright points of galaxies to reveal how these larger-scale shapes were beginning to form. First by twos and threes, then by dozens and hundreds, clusters of galaxies began to form. This clustering was speeded up by the additional gravity provided by dark matter. Without dark matter, the number of clusters of galaxies and their larger-scale filamentary patterns would be dramatically smaller.

Below *Located in the constellation of Hercules, about 230 million light-years away, NGC 6052 is a pair of galaxies that are in the process of colliding. Eventually, the galaxies will fully merge to form a single, stable galaxy.*

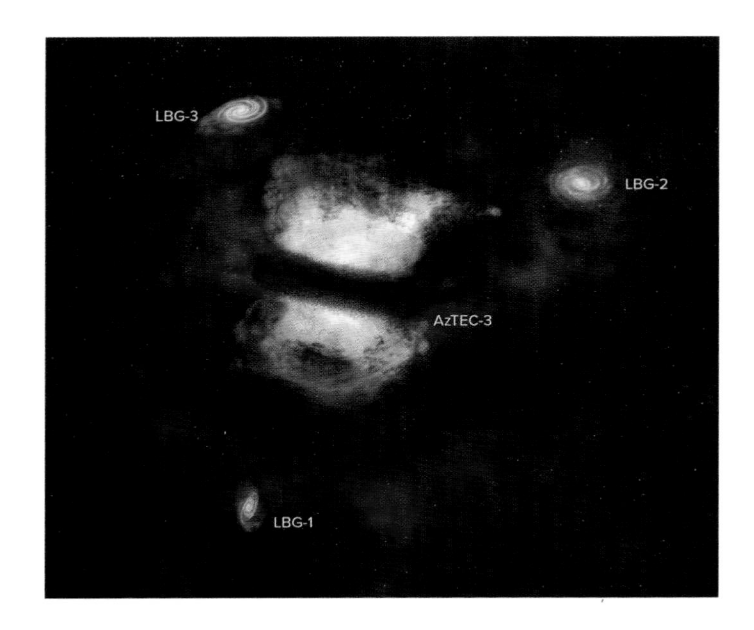

Left *Artist's impression of the proto-cluster AzTEC-3, along with three smaller, less active galaxies. AzTEC-3 recently merged with another young galaxy and the whole system represents the first steps toward forming a galaxy cluster.*

Below, left *Stephan's Quintet is a group of galaxies 300 million light-years away, gravitationally interacting with one another and causing distortions in their shapes. The interactions have produced bursts of new star formation. The scale of the image is about 500,000 light-years across.*

Below, right *Four galaxies located 1 billion light-years from Earth are in the process of collision and merger in the center of the picture in a region about 30,000 light-years across. The resulting infrared source called IRAS 19297-0406 implies that the collision is producing about 200 new, massive stars every year—about 100 times more stars than our Milky Way creates.*

One of the youngest known clusters of galaxies called AzTEC-3 is in the process of forming at an age of about 1.1 billion years ABB. This object consists of five smaller galaxy-like clumps of matter, each forming stars at a prodigious rate. We now begin to see how some of the small clumps in this cluster are falling together and interacting, eventually to become a larger galaxy-sized system. Astronomers estimate that by 2.2 billion years after the Big Bang, half of all the massive elliptical galaxies we see around us today had already formed from these kinds of merger events.

Although the formation of galaxy clusters in the early history of the universe cannot be easily discerned even with the most powerful telescopes, astronomers have many examples of this process to study at high resolution in the nearby universe. One of the most dramatic of these is called Stephan's Quintet. Other objects such as NGC 6052 and IRAS 19297-0406 reveal how the collision between galaxies can build up even larger systems. Over the course of billions of years and hundreds of collisions, small dwarf galaxies can grow into vast Milky Way-sized galaxies that are common by the current age of the universe.

TYPES OF GALAXIES IN THE MODERN UNIVERSE

Elliptical galaxies are by far among the most numerous and simple. They appear as round balls of stars, but often contain billions of very old "Population II" stars that formed in the universe once the still-older Population III stars had vanished. Some, such as the nearby dwarf elliptical galaxy Messier 32 in Andromeda is only 7,000 light-years in diameter and contains only about three billion old, red stars. There are no signs of new stars being born. It does, however, have a supermassive black hole at its core that is about one-third the mass of the one in our Milky Way's core. There are also giant elliptical galaxies, such as Messier 87 in the Virgo galaxy cluster, which is as big as our Milky Way and has a mass of over two trillion stars. Its core contains a supermassive black hole, but this one is incredibly active making M87 one of the most powerful radio sources in the sky next to our sun.

Above *The Abell S0740 galaxy cluster is more than 450 million light-years away in the direction of the constellation Centaurus. The giant elliptical galaxy ESO 325-G004 looms large at the cluster's center. The galaxy is as massive as 100 billion of our suns.*

Spiral galaxies are among the prettiest galaxies in the universe. They have a central nucleus of stars from which arms filled with stars and nebulae radiate in a curved, pinwheel shape. The smallest known spiral galaxy, NGC 5949 is barely 30,000 light-years across with perhaps no more than 10 billion stars. Then there are gargantuan spirals such as NGC 6872, which is five times the diameter of our Milky Way and contains more than a trillion stars. Spirals are sometimes found in association with galaxies that are gravitationally interacting, so that astronomers believe that some spiral galaxies get their dramatic arms from these encounters. Unlike the elliptical galaxies that have little or no interstellar gas and dust to form new stars, spiral galaxies have up to 20 percent of their mass in interstellar clouds. This allows new stars to form steadily over billions of years. The most massive of these stars and nebulae help to define the shapes of the spiral arms.

Above *Messier 74—the perfect spiral galaxy located 30 million light-years from the sun is about the same size and mass as our Milky Way.*

Barred spirals are an unusual class of spirals and our own Milky Way is one of these. The Milky Way is a mid-sized barred spiral with a diameter of 100,000 light-years and a mass of up to two trillion solar masses, including dark matter. Like spirals, they share the same sizes and masses and types of stars, but, in their central regions, stars are not orbiting the core in circular or elliptical paths. Instead, the orbits of the gas and stars have been reshaped by a resonance effect into a bar-like pattern, and then maintained through gravitational interactions.

Above *Barred spiral galaxy NGC 1300 photographed by the Hubble telescope is located 61 million light-years from the sun, and, with a diameter of 110,000 light-years, is slightly larger than the Milky Way.*

Irregular galaxies are just what their names suggest. They have no specific geometric shape, although some may have traces of arms or bright core regions. Typically, these are filled with a mixture of very young and very old stars like spirals. The Large and Small Magellanic Clouds visible in the Southern Hemisphere are spectacular examples in our neighborhood. Irregular galaxies can have masses from 1 million to 10 billion stars and are typically about one-tenth the size of the Milky Way. They may have played an important role in forming the first generations of stars and galaxies because they are easily torn apart by the gravity of other galaxies, which can then use their stars and gas to grow larger in time.

Above *The irregular galaxy NGC 4449 is located only 12 million light-years from the sun and is about one-tenth the size and mass of the Milky Way.*

Peculiar galaxies are a catch-all category for galaxies that seem as their name suggests peculiar in shape or some other aspect. Very often, their odd shapes have resulted from recent collisions with other galaxies that in some cases can still be seen in progress. Peculiar galaxies are often the hosts for enormous bursts of star formation activity, supernova activity or even the outpouring of energy from central supermassive black holes.

Above *The two interacting galaxies in Arp 273, located 300 million light-years from the sun, are an example of peculiar galaxies that do not look like "normal" galaxies. Gravitational distortions as well as intense bursts of star-forming activity make these galaxies unique in their shapes.*

// Active galactic nuclei

During the first few billion years ABB galaxies are still being formed. The most common way they grow is by cannibalism and collision. These collisions are violent and, like gasoline dumped on a fire, some of the material floods the central cores of these galaxies where massive black holes consume the matter and emit enormous quantities of energy. These galaxies are called quasars. Although they were discovered in the 1960s, it took another 30 years before the advent of the Hubble Space Telescope allowed astronomers to look at the images of quasars to see what they actually looked like. What astronomers found was that quasars almost always seem to be associated with the dense core regions of galaxies, and that these galaxies were usually interacting with their neighbors during collision and cannibalism events. Somehow, the "monster" inside the nuclei of some galaxies was being fed huge amounts of stars and gas, which released the quasar light.

Careful studies of the distances to thousands of quasars reveals that these objects formed very rapidly once the universe became a few billion years old. This quasar era has lasted up to the present time. Supermassive black holes, formed during this era, consumed all of the material surrounding them in violent spurts of growth, and then became fainter as their fuel supply ran out so they could no longer be detected. The feeding process can, however, be re-awakened if new material starts to flow into the black hole. This can happen during times when galaxies collide, which places gas and stars on orbits that can intersect the central black holes. Nearby galaxies that are continuing to feed their central supermassive black holes have been recognized since before black holes were, themselves, considered as possible energy sources. Many different types of these "active galaxies" have been discovered and classified into separate categories depending on their appearances. Galaxies with bright central nuclei but with signs of tremendous gas speeds were called Seyfert galaxies, while galaxies with a bright core that varied rapidly in time were called BL Lactertae (or BL Lac) objects, because astronomers thought that they were merely a form of variable star within our Milky Way. The advent of radio astronomy identified galaxies that seemed normal looking but were titanic sources of radio energy, often mapped into two lobes located well outside the "radio galaxy" itself.

During the 1980s and 90s, astronomers began to realize that active galactic nuclei could actually be a single phenomenon viewed from different orientations in space. A supermassive black hole actively feeding from its dusty accretion disk would resemble a Seyfert galaxy if viewed edge-on so that the dust obscured any details except the high-speed gas motions. The disk near the black hole would be ejecting gas at high speeds close to the speed of light. When viewed along the axis of these jets, you would only see the blinding light from the jets flowing toward you. This material would be lumpy and produce the rapid brightness changes seen in BL Lac objects. Finally, once this high-speed plasma leaves the galaxy, it collects in vast reservoirs of plasma that are detected by their radio emission, producing the classic radio galaxy shapes.

Below *Quasar host galaxies revealing that the source of the quasar emission comes from a small region within the core of these galaxies, whose spiral arms can be discerned.*

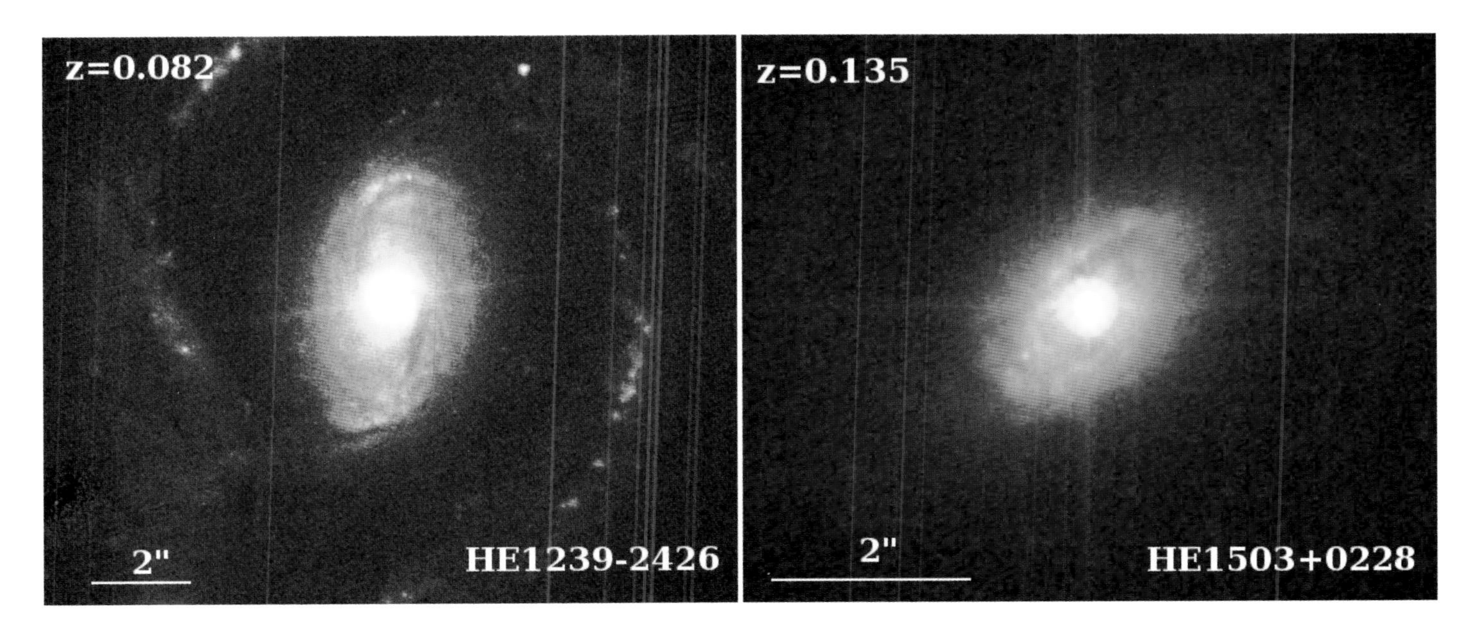

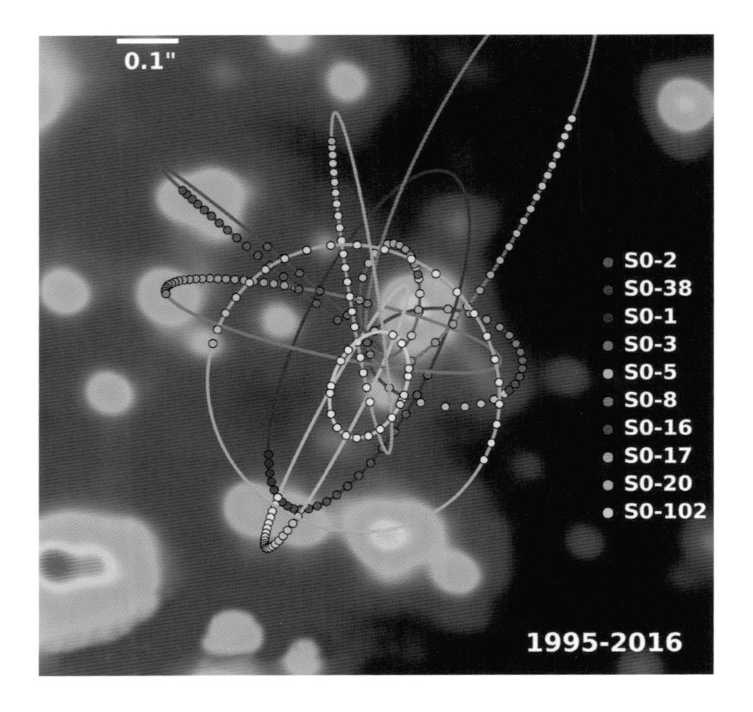

Above *NGC 7742, an example of a Seyfert galaxy with an active core region but emitting less energy than a quasar.*

Left *The orbits of the stars located near the Milky Way's supermassive black hole, called Sgr A*, reveal that it has a mass of four million times our sun's mass, and is located at the common foci of the elliptical orbits.*

Below *Spectacular jets powered by the gravitational energy of a supermassive black hole in the core of the elliptical galaxy Hercules A.*

Galactic superclusters and local cosmic geography

After 50 years of discovering our galactic neighbors in space, we finally have a good idea of what our neighborhood looks like. Our Milky Way and the Andromeda galaxy form a pair of massive spiral galaxies traveling through the universe. We are not traveling alone but are accompanied by a retinue of 80 other, smaller "dwarf" galaxies that are gravitationally attracted to either the Milky Way or Andromeda. This collection of galaxies, called the Local Group is about ten million light-years across. These two galaxies are unusually large; each has about 300 billion stars, while the dwarf galaxies contain fewer than a few billion each. Unfortunately, Andromeda and the Milky Way are moving toward each other at about one light-year every 200 years. In another four billion years they will collide. Before then, many of the dwarf galaxies may finally be consumed by either Andromeda or the Milky Way.

At an even larger scale, our Local Group is itself surrounded by other individual galaxies and groups of galaxies. The next-largest collection is called the Virgo Cluster. It contains as many as 2,000 galaxies and represents a huge amount of mass; almost 100 trillion times

Below *The galaxies in the Local Group form a binary system consisting of the two massive galaxies: Andromeda and the Milky Way, along with dozens of other galaxies that are gravitationally bound to them. They travel through space as a family of objects*

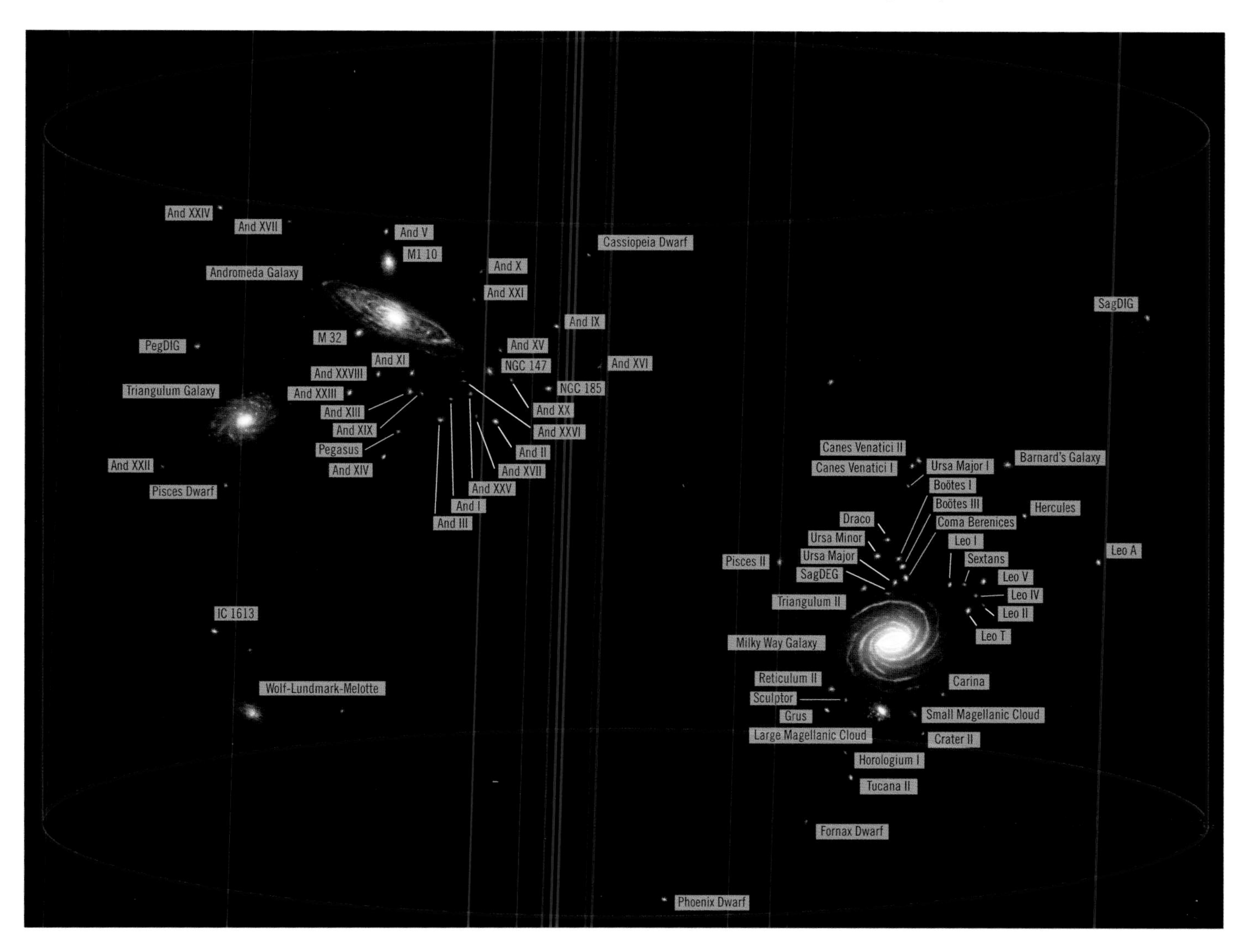

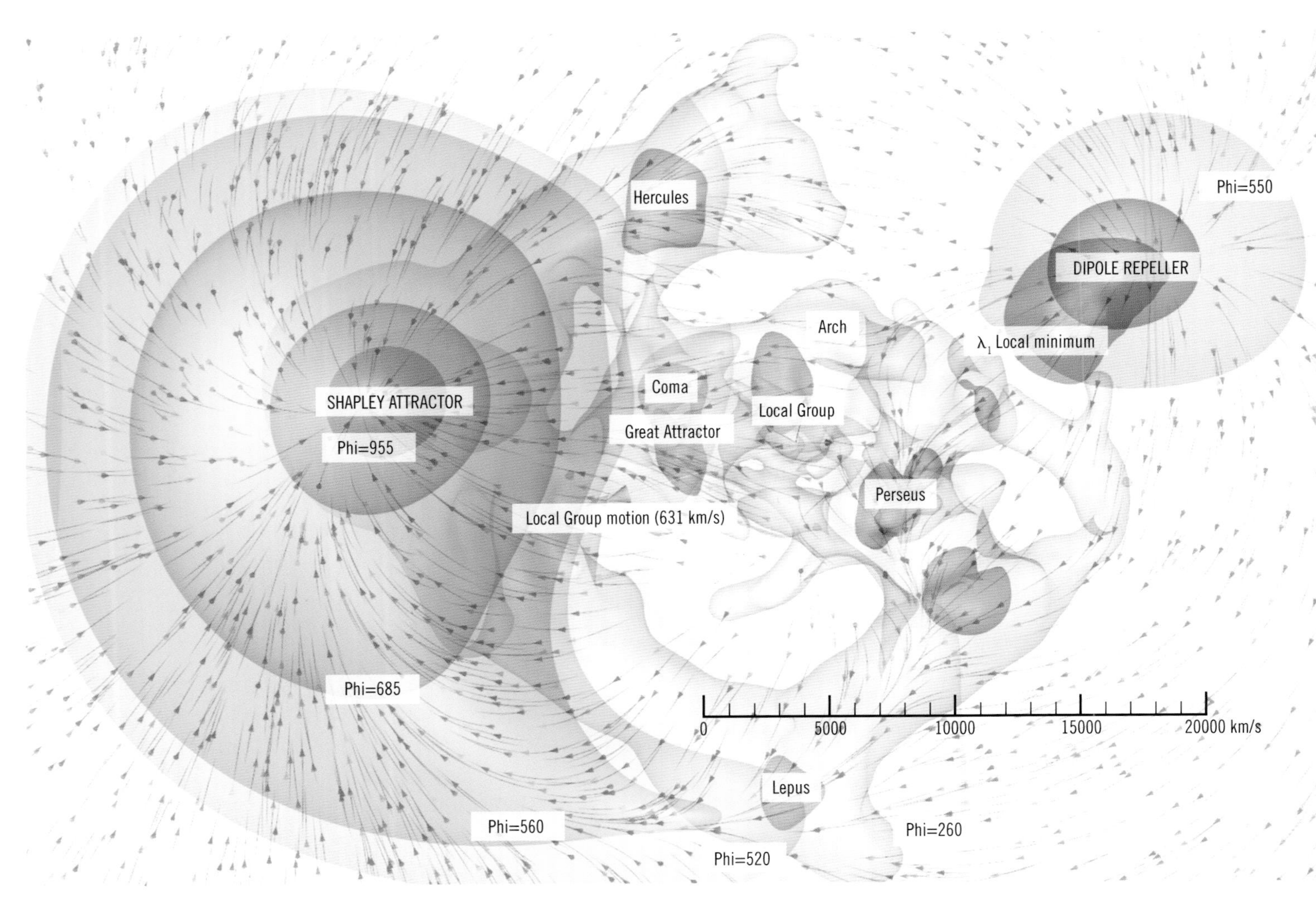

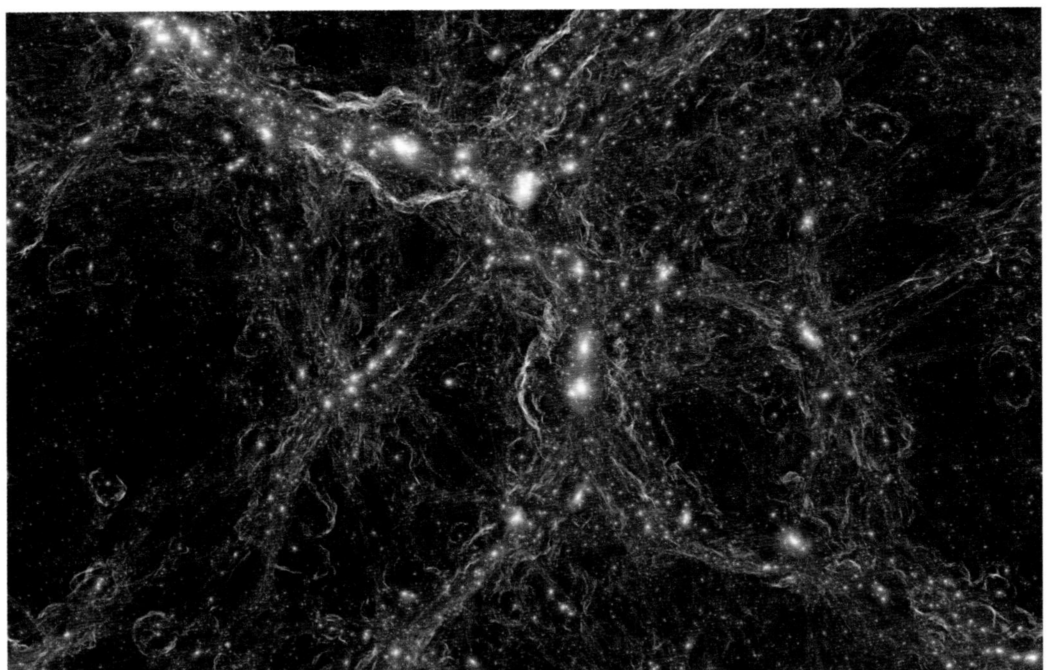

Above *The movement of the Laniakea Supercluster (gray) toward the Shapley Supercluster (green). The lines with arrows show the directions of motion of galaxies.*

Left *A visualization of the Laniakea Supercluster, which represents a collection of more than 100,000 estimated galaxies spanning a volume of over 300 million light-years, shows the distribution of dark matter (shadowy purple) and individual galaxies (bright orange/yellow) together. Courtesy of TNG Collaboration.*

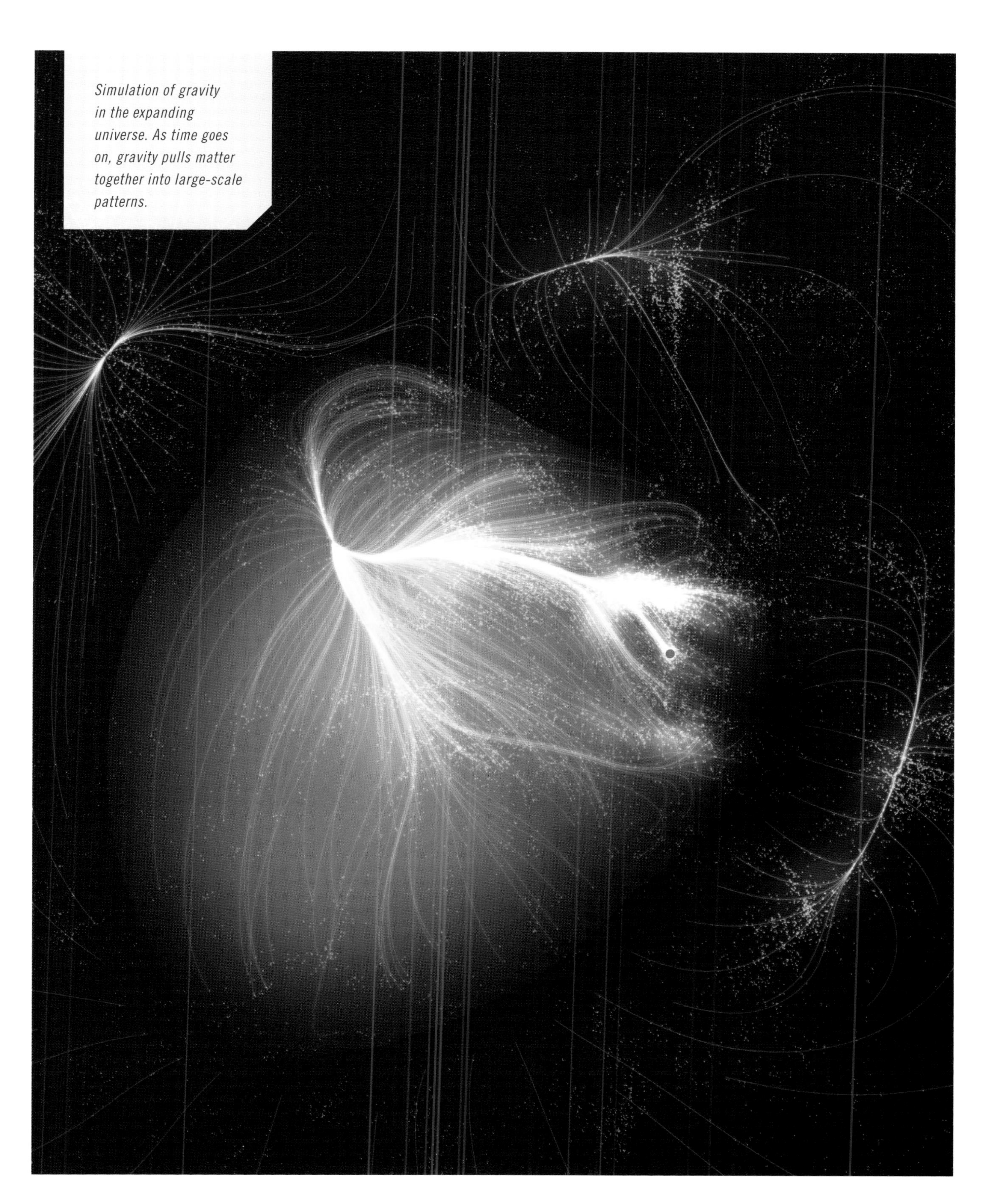

Simulation of gravity in the expanding universe. As time goes on, gravity pulls matter together into large-scale patterns.

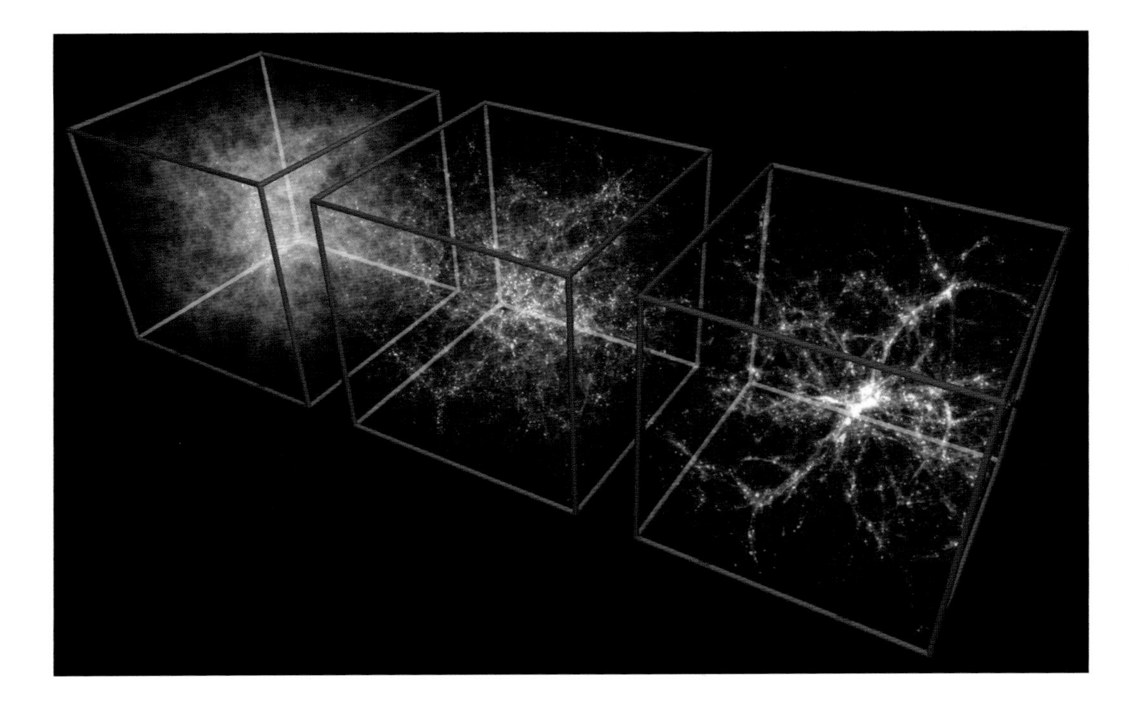

Left *Astronomers have discovered that the Milky Way galaxy is part of a supercluster which they've named Laniakea. The Laniakea Supercluster encompasses 13 major galaxy clusters, including the Virgo Cluster. Galaxies are not distributed randomly through the universe, but lie in clusters and thin filaments lining the edges of gigantic empty voids. Galactic superclusters are formed when the thin filaments intersect.*

the mass of our sun, located about 54 million light-years from the Milky Way. The Local Group feels the gravitational influence of the Virgo Cluster and its motion through the universe is headed for the center of the Virgo Cluster. In fact, the Virgo Cluster is so massive that it has also altered the motions of a hundred other clusters of galaxies like our Local Group as far away as 60 million light-years from its center. Astronomers refer to this ensemble of more than 10,000 galaxies as the Virgo Supercluster. The Virgo Supercluster is believed to be only one of an estimated 10 million other superclusters in our observable universe.

Superclusters are vast collections of individual clusters of galaxies but are by no means the largest objects that exist in our corner of the universe that have been discovered through cosmic cartography. The process of cosmic cartography is tedious. It is not enough to merely note where each of millions of galaxies are located across the sky as viewed from Earth. You also need to determine their distances and speeds of motion. Distances can be determined from one of several different methods, but for the speed of the galaxy through space, that is a serious challenge.

The movement of an object through three-dimensional space requires measuring its speed along each of the three "axes." Like a police officer using a radar gun measuring a car's speed, astronomers can measure the speed of a galaxy along the direction toward and away from Earth using the Doppler Effect. But galaxies are so far away astronomers can never measure their side-to-side movement in the other two dimensions. So, to build a model of where galaxies are in space, how they are moving, and which objects are affecting them gravitationally, astronomers have to use a supercomputer and Newton's law of universal gravitation. They make a model of where the galaxies appear to be in space as viewed from Earth, estimate their masses from careful observations of them as individuals, and then predict what the measured speed should be viewed from Earth using the Doppler Effect. They tweak the locations and masses of the galaxies until, after running millions of these models, they arrive at ones that closely resembles the observations.

These models not only give you the three-dimensional locations of each of the galaxies in space, but also tell you in which directions they are moving through intergalactic space. When this is done, the Virgo Supercluster is found to be only one of a dozen other superclusters within several hundred million light-years of the Milky Way that are being gravitationally attracted toward a large collection of galaxies called the Shapley Supercluster. The center of this enormous mass is located 650 million light-years from the Local Group, and situated in the direction of the constellation Centaurus. Our Virgo Supercluster, along with a dozen other superclusters, has been named the Laniakea Supercluster after the Hawaiian word for "immense heaven."

Because there has not been enough time in the current age of the universe for a single galaxy to cross the scale of its own cluster of galaxies, most clusters and superclusters are not actually bound systems at all, but merely objects temporarily passing in the night and probably destined to dissolve in the far future.

THE MILKY WAY

One of the most dazzling sights in the evening sky is the great swath of the Milky Way crossing the sky from horizon to horizon. A staple of mythology and admired for millennia, only recently have we been able to gauge its true scope and shape. It is a delicate pinwheel of stars and clouds spinning majestically through the dark intergalactic spaces of our universe. Its center hosts a massive and forbidding black hole, while our sun and its stellar neighbors orbit safely every 250 million years at a distance of 27,000 light-years.

The Milky Way has transfixed the human imagination for millennia. From any dark sky location far from city lights, this band of dim light is filled with numerous star clouds and nebulae.

// Basic contents

A careful study of the night sky reveals not only a bewildering number of stars forming the basic constellations, but smaller objects such as star clusters and nebulae. The most famous of these star clusters is the Pleiades, which was known to the Ancient Greeks who named its seven stars after the daughters of Atlas. One of the earliest known depictions of the Pleiades is found on the Nebra sky disk from the Northern German Bronze Age dated to about 1600 BCE. There are also unresolvable blotches of light called nebulae, of which the most famous is the Orion Nebula, first recorded in 1610 soon after the telescope became popular. The idea that the Milky Way consisted of more than just stars didn't really materialize until astronomers began serious telescopic studies of the sky. The first catalog of nebulae to be avoided if you were a comet hunter was created by Charles Messier in the 1760s using a small 4 inch (10 centimeter) refractor. His first object, "Messier 1," was the Crab Nebula in the constellation Taurus, which later turned out to be a supernova remnant. Other entries in his 110-object list included star-forming nebulae such as Messier 42 (the Orion Nebula), and star clusters such as Messier 45 (the Pleiades), as well as galaxies beyond the Milky Way such as Messier 32 (the Andromeda galaxy). When William Herschel between 1786 and 1802 published his *Catalogue of Nebulae and Star Clusters* using his 12 and 18-inch (30- and 45-centimeter) telescopes, not only did it include Messier's objects, but it had over 2,500 additional entries. What all of these objects were was a matter of speculation, especially since no one had any good way to determine their distances, let alone the distances to the stars themselves. There were some very basic things that you could deduce from how they were distributed in the sky.

The small, faint "extragalactic nebulae" tended to be found above and below the band of light in the sky of the Milky Way. Other bright and irregular nebulae such as the Great Nebula in Orion tended to be found inside the Milky Way band, while the small numbers of roundish "planetary nebulae" seemed to be both inside and outside this band but did not look at all like the fainter extragalactic nebulae that were much further from the plane of the Milky Way. Today, astronomers recognize a small number of basic kinds of objects that make up the Milky Way galaxy.

Two kinds of star clusters exist: galactic and globular. Galactic star clusters such as the Pleiades are irregular collections of dozens or hundreds of stars born from the same cloud of gas, traveling together as a group in space close to the plane of the Milky Way. Globular clusters are round systems of up to one million stars that are found high above the plane of the Milky Way, and mostly located toward the constellation of Sagittarius.

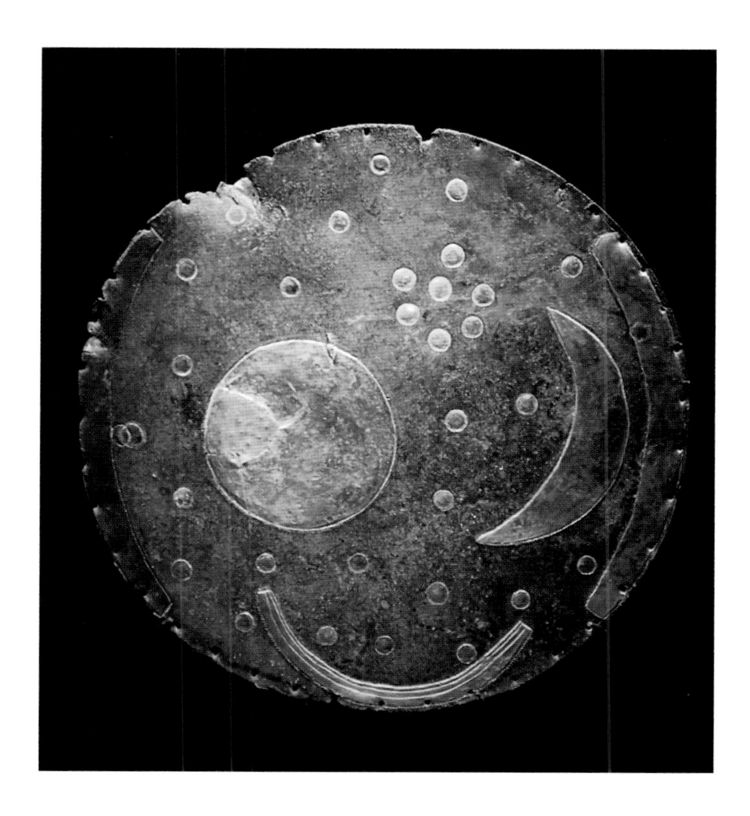

Above *A depiction of the Pleiades cluster, the sun and the moon on the Nebra sky disk.*

Nebulae are very diverse in their shapes and come in two types: dark nebulae and bright nebulae. Dark nebulae, such as the Horsehead Nebula in Orion, are dark clouds of gas seen against the field of bright background stars or nebulosity. These interstellar clouds have so much dust that they block the light from background sources. Bright nebulae come in three major classes: supernova remnants, planetary nebulae and emission nebulae. The supernova remnants are often round in shape such as the great Cygnus Loop, but can also be irregular like the Crab Nebula. They are illuminated by intense ultraviolet and X-ray light produced by the central exploding star whose dispersing material makes up the nebula. Planetary nebula, such as the Ring Nebula in Lyra, are generally round but can also have complex bow-tie or bipolar shapes. These are produced when stars like our sun leave their red giant phase and their outer layers are ejected into space leaving behind a white dwarf. Emission nebulae, such as the Great Nebula in Orion, also called HII (H-two) regions, are

Above *Photograph of the Milky Way obtained by an ordinary cell phone camera.*

produced by young massive stars whose ultraviolet light is ionizing the interstellar gases that surround them. Supernova remnants, dark nebulae, and emission nebulae are all associated with the formation of young stars and so are very common within the disk of the Milky Way. Planetary nebulae are produced by evolved older stars and can be found thousands of light-years above or below the main disk of the Milky Way. Globular clusters, of which there are fewer than 160, are satellites to our Milky Way and are found in, and define, what is called the halo of the Milky Way.

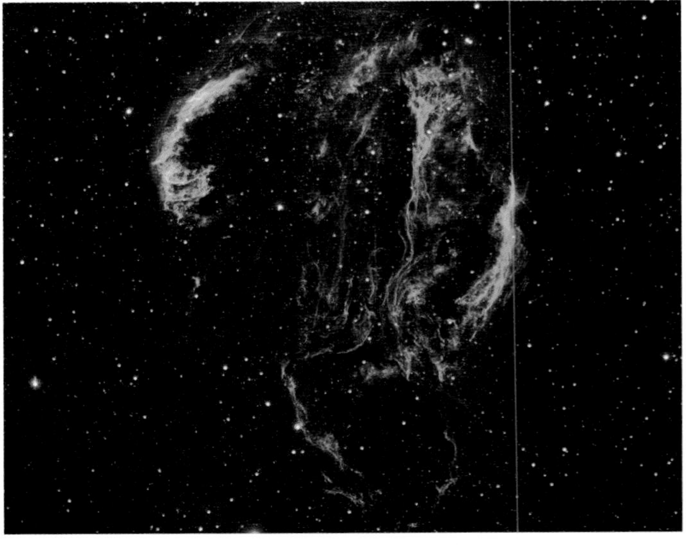

Right *The Cygnus Loop nebula is all that remains of a nearby supernova explosion that occurred 21,000 years ago when the last Neanderthal had vanished and* Homo sapiens *became our ascendent ancestor.*

// Exploring its structure from the inside out

For thousands of years scientists thought that our universe with all of its stars was spherical in shape because that was the most perfect shape for a celestial object. But the band of diffuse light across the sky called the Milky Way was soon resolved by Galileo Galilei's telescopic observations in the early 1600s, revealing that it was not just some misty, cloudy celestial gas but uncounted millions of individual stars combining their light across space.

By the late 1700s the British astronomer Sir William Herschel completed his star-gauging research to map out the shape of the Milky Way. It was a tedious process of breaking up the sky into a geometric gridwork and then, at the eyepiece of his telescope, literally counting the numbers of stars in each grid. When he was finished, he published the first astronomical map of what the Milky Way looked like from the outside: It was a fish-shaped blob of

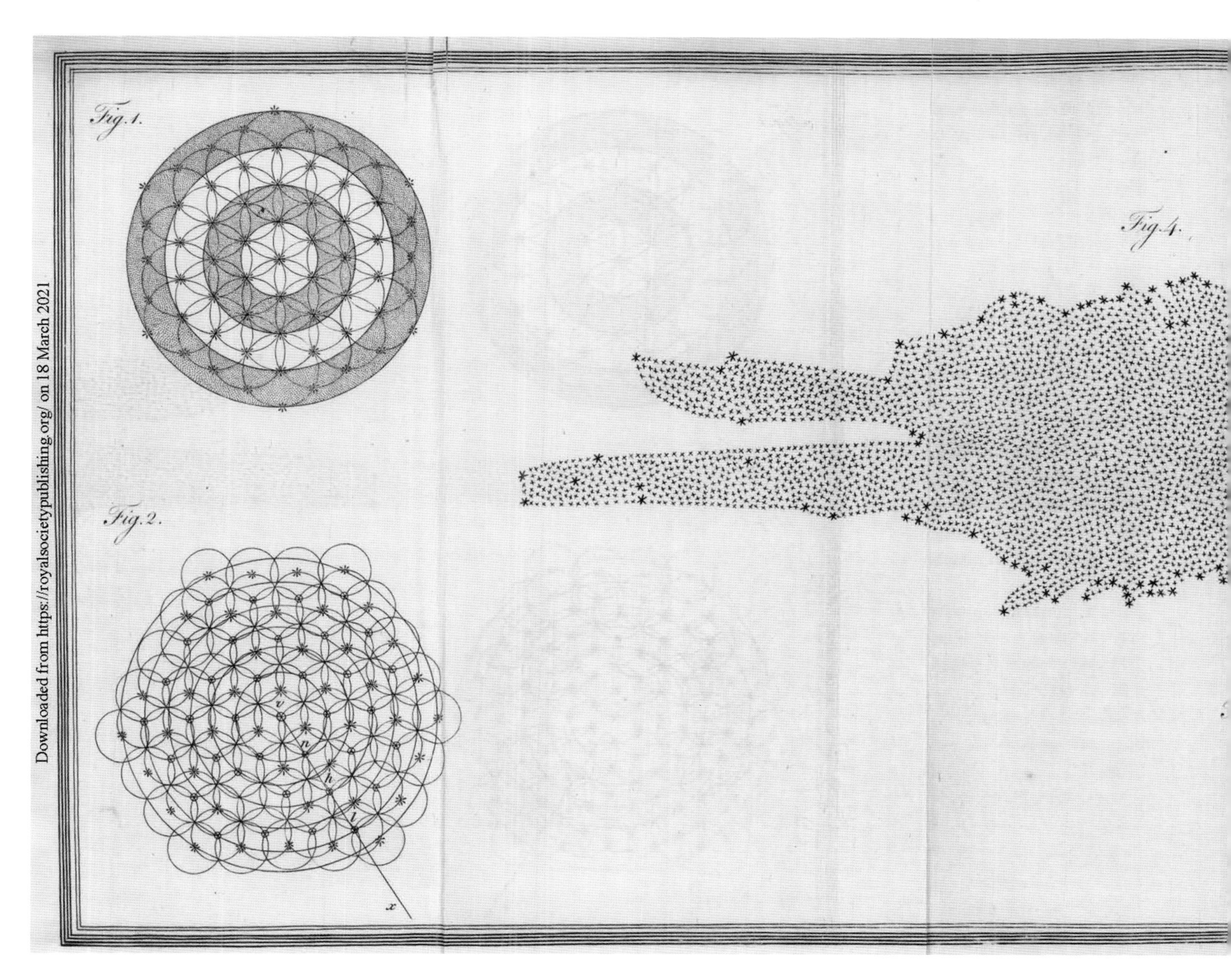

stars with several major tendrils, and with most of its stars clustered in the sky towards the constellation Sagittarius. In his provocatively-titled book *On the Construction of the Heavens* in 1785, he called it "A very extensive, branching, compound Congeries of many millions of stars."

William Herschel and other astronomers afterward had also studied a variety of nebulae in the sky, seen as smudges of light, but some revealed disk or spiral-like shapes. One of these, the Whirlpool Nebula in the constellation Canes Venatici was sketched in detail by Lord Rosse using his enormous telescope in 1845–50 revealing a central nucleus surrounded by misty arms.

It didn't take too long for astronomers to realize that there were three common shapes for these nebulae: amorphous blobs of light; spirals; and objects with roundish, elliptical shapes. A quick look at the night sky eliminated all but the flat spiral forms because, for the shapes, our sky from the inside of these objects would be filled in all directions by stars, not just along the narrow band of the Milky Way. In a leap of insight, the Dutch journalist and astronomer Cornelius Easton *c.*1900 thought that our Milky Way might resemble one of the spiral nebulae and so translated this idea into a perspective of what such a system would look like from sky observations made from Earth.

Then, in 1915, the American astronomer Harlow Shapley worked out the distances to dozens of globular clusters, which he thought were satellites to the Milky Way and found that, whatever its shape might be, most of its mass was concentrated about 30,000 light-years in the direction of the constellation Sagittarius.

One of the confusing issues encountered by early astronomers was the intrinsic brightness of a star. If you take a 100-watt bulb and place it at different distances it appears brighter or fainter. It is a simple matter to relate its distance, via the inverse-square law, to its apparent brightness so faint stars in the sky would be further away than the bright ones. The problem is that stars are not all "100-watt bulbs." Astronomers were not aware until the 20th century that they vary enormously in their wattages, and so you can't directly relate how they appear in the sky to a distance to them. For example, the naked eye star Deneb

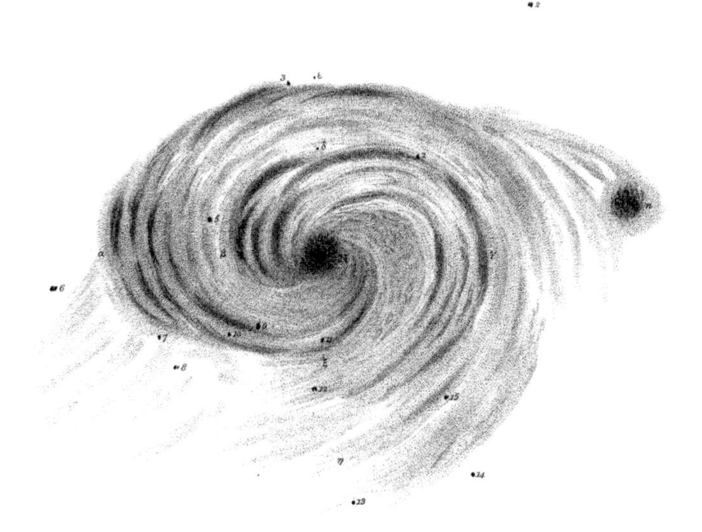

Above *First sketch of the Whirlpool Nebula (Messier 51) by Lord Rosse.*

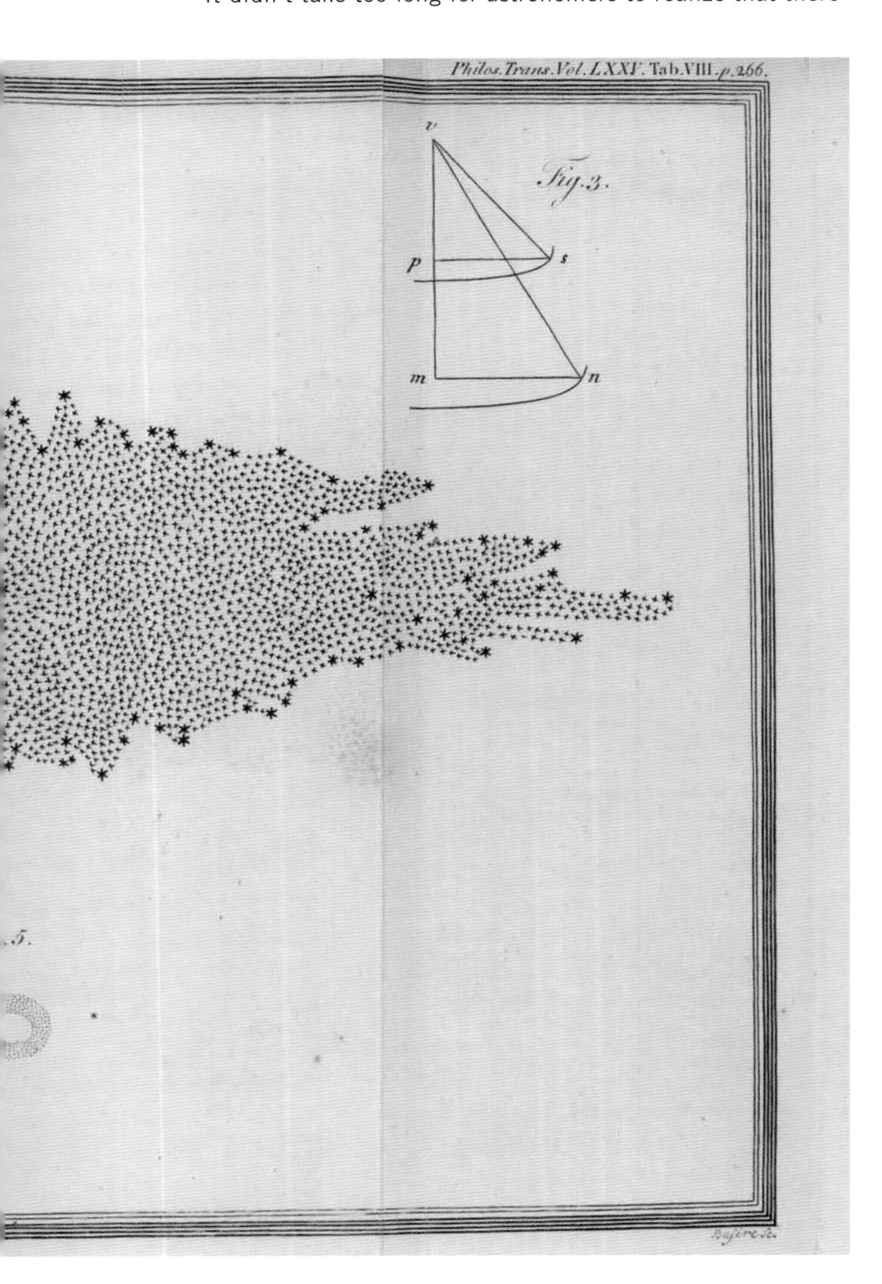

Left *Herschel's map of the Milky Way was the product of thousands of hours of "star gauging" at the eyepiece. The odd shape arises from assuming that all stars have exactly the same brightness and that their light is not obscured by interstellar dust.*

in the constellation Cygnus appears just as bright as the star Regulus in the constellation Leo, but Deneb is 196,000 times as luminous as our sun while Regulus is only 288 times the sun's wattage. Deneb is located 2,600 light-years from the sun while Regulus is close by at only about 80 light-years. When Herschel created his star map of the Milky Way, such luminosity differences were not as yet suspected because stellar distances could not be determined. This resulted in a map of the Milky Way that was foreshortened in some directions and expanded in others.

Below *The globular cluster NGC 6397 located at a distance of 7,800 light-years reveals many red and blue stars clustered in a tight core only a few light-years across. At an age of 13.4 billion years, it was formed soon after the Big Bang and now contains very old red and blue giant stars. The blue stars are much younger and were created by collisions between some of the older stars.*

Another feature of the Milky Way that was not suspected until 1930 when astronomer Robert Trumpler announced that interstellar space contained dust grains capable of dimming starlight. In some cases such as the Horsehead Nebula in Orion, this dust in the form of clouds can dramatically alter the shape of luminous nebulae that they are often adjacent to. Interstellar dust is present everywhere, and amounts to dimming the brightness of stars by about a factor of five for every 3,000 light-years of distance. Herschel's Milky Way map based on the apparent brightness of stars, along with the Dutch astronomer Jacobus Kapteyn's improvement of it in 1922, making it appear more elliptical, did not include the dimming of starlight by interstellar dust and so this led to models that could give estimates of 30,000 light-years in length and 6,000 light-years in thickness, but could not account for the actual scale of the Milky Way implied by Harlow Shapley's globular cluster distances.

The Horsehead Nebula in the constellation Orion is a vivid example of a dark foreground dust cloud blotting out the light from more distant stars and luminous nebulae behind it.

// Star populations

One of the first things astronomers learned was that our Milky Way star system actually contains many different populations of stars—each with their own history dating all the way back to the Big Bang itself.

Our galaxy is embedded in a vast halo of stars stretching out at least 500,000 light-years from its core regions. These Population II stars have very few heavy elements such as iron or oxygen and were probably among the first stars to be born in our galaxy some 13 billion years ago. Globular clusters are filled with these "metal-poor" stars. Next, we have a population of stars in the nuclear regions of the Milky Way that have been slightly more enriched with heavier elements. Together, the nuclear and halo populations are the oldest in our Milky Way, but they are embedded within a vaster dark matter halo that contains about ten times more mass than what we see in the luminous stars, though this dark matter is utterly invisible except for its gravitational influences.

As we explore the distant universe to gather images of the youngest galaxies as they formed over 12 billion years ago, we can also turn to our own Milky Way to see this process from the inside-out. Apparently, the halo of our Milky Way accumulated first from material that was already slightly enriched by the long-vanished, massive Population III stars that went supernovae. The Population II stars formed next, and in fact several of these old stars have actually been dated. The star called J0815+4729 is located 7,500 light-years from the sun in the halo of the Milky Way and likely formed just 300 million years after the Big Bang, some 13.5 billion years ago. Astronomers have obtained ages for eight of these "Methuselah" stars older than about 13.2 billion years. In comparison, stars found among the 150 globular star clusters that orbit the Milky Way may be slightly younger, and many were formed around 12 billion years ago.

Within this spherical ensemble of older stars and globular clusters there is a flattened disk of stars that are perhaps 10 billion years old or less, and contain even more heavy-element enrichment. These Population I stars have masses similar to, or a bit less, than our sun and up to 3 percent of their mass is in elements heavier than hydrogen and helium. Many of them have reached the end of their lifetimes after 12 billion years, and are evolving into planetary nebulae with white dwarf cores. When astronomers map out the locations of planetary nebulae in the sky, the pattern they get in space

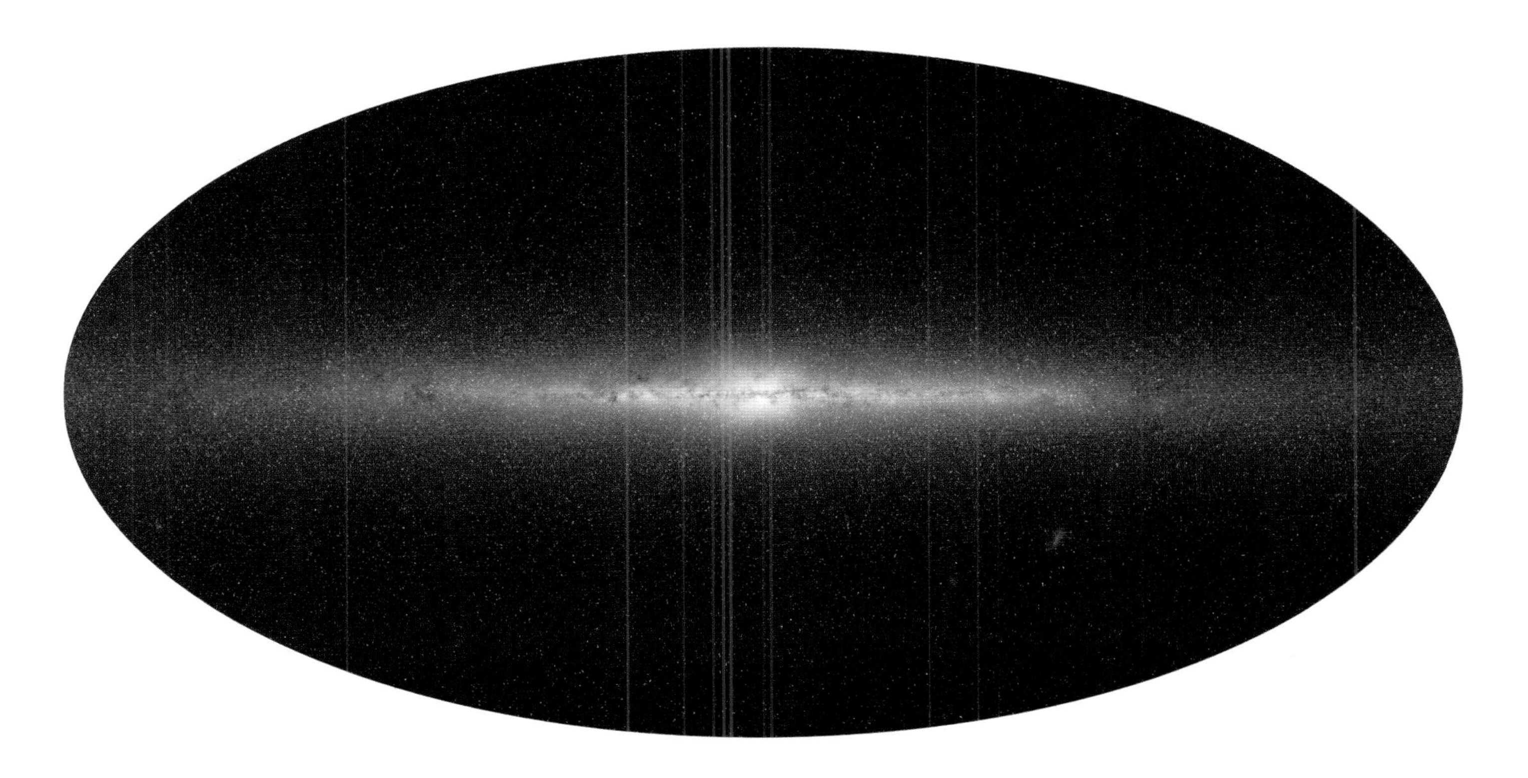

Below: *The stars in the Milky Way can be classified into several distinct groups, including an older halo population, a thick disc population, and a thin disc population. Each population arose from separate evolutionary events as the Milky Way formed over 12 billion years ago.*

Above: *This globular cluster, called Messier 80 (NGC 6093), is one of the densest of the 147 known globular star clusters in the Milky Way galaxy. At a distance of 26,000 light-years it has an age of about 12 billion years.*

follows this "thick disk" population of stars and are spread out within 1,000 light-years of the disk of the Milky Way.

Finally, the youngest Extreme Population I stars less than a few billion years old are found in an even narrower thin disk only a few hundred light-years thick. Many of the most massive stars in the Milky Way are found in this zone and are still close by the vestigial gas and dust clouds from which they formed. These stars are metal-rich and in some cases possess even more heavy elements than our own sun. The more massive members of this population erupt in supernovae explosions. Millions of these have occurred since the formation of our Milky Way and have littered interstellar

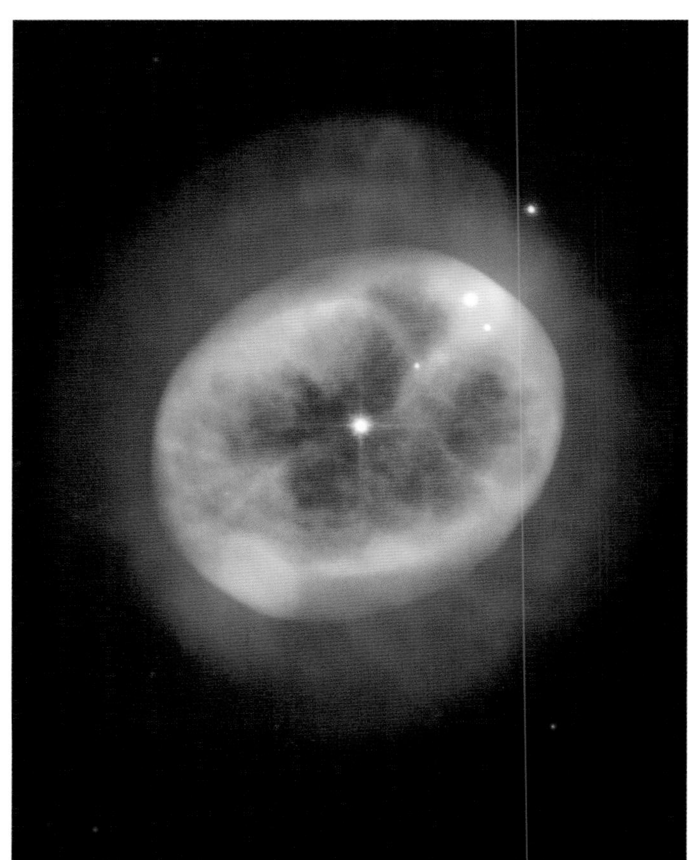

Right: *The planetary nebula NGC 2022 located 8,200 light-years from the sun. It is the remains of a star that had slightly more mass than our own sun.*

The Orion Nebula and its associated cluster of young massive stars is less than 5 million years old. This object is one of the closest stellar nurseries for both low- and high-mass stars, at a distance of about 1,350 light-years.

Above *This image shows the dark cloud Barnard 68 located 400 light-years from the sun in the constellation Ophiuchus. At visible-light wavelengths, the small cloud is completely opaque because of the obscuring effect of dust particles in its interior. These clouds are very common in the disk of the Milky Way.*

space with gases rich in heavy elements.

The young stars in the disk of the Milky Way are often found close by the dense gas clouds from which they formed, and that also define the location of the spiral arms of the galaxy. Some of these clouds are near the sun at distances of only a few hundred light-years and can be seen blotting out the light from more distant stars. These dark clouds are up to several million times as dense as ordinary interstellar hydrogen clouds (about one hydrogen atom per cubic centimeter) and provide the matter from which stars are formed in their interiors. Typically, more than one star forms at a time before the cloud dissipates, leaving behind collections of stars called galactic or "open" star clusters.

Left *SN 1006 supernova remnant. Massive stars explode and scatter elements heavier than helium into the interstellar medium from which stars are then born and become enriched with heavy elements.*

// The shape of the Milky Way

The advent of radio astronomy by Karl Jansky in 1933, and the later work in the 1940s by astronomer Jan Oort in the Netherlands and Edwin Purcell in the United States, eventually allowed astronomers to detect clouds of hydrogen gas just about anywhere within the confines of the Milky Way. The speeds and directions of these clouds were then used to map out their locations in space, revealing a pinwheel-shaped flattened system of gas clouds with several distinct arms. It was pretty clear by the late 1950s that our galaxy was indeed some kind of spiral galaxy, but most of its structure was hidden by dense clouds of dust that blocked starlight from more than a few thousand light-years from the sun.

An interesting feature of our Milky Way is that it is rotating in space. The specific speed depends on your distance from its central mass, just as the planets in our solar system travel slower the further from the sun they are. This is not a random decrease in orbit or rotational speed but is carefully choreographed by the amount of mass interior to the orbit according to Newton's law of universal gravitation. The result is called differential rotation because the speeds decrease the further from the nucleus you go. But over time, radio astronomers using measurements of the speeds of hydrogen gas clouds discovered an inconsistency.

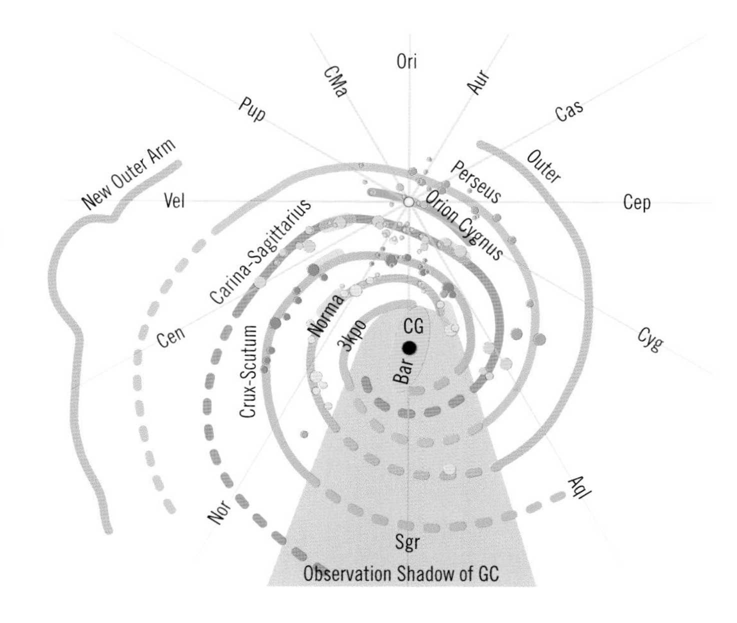

Above *In this diagram with the sun represented as the yellow dot at the top, the spiral arms of the Milky Way are colored differently in order to highlight what structure belongs to which arm.*

Below *One of the first maps of the interstellar hydrogen in the Milky Way by radio astronomers reveals a definite spiral arm pattern. Our sun is at the center of the cross with the center of the Milky Way at the point "C+."*

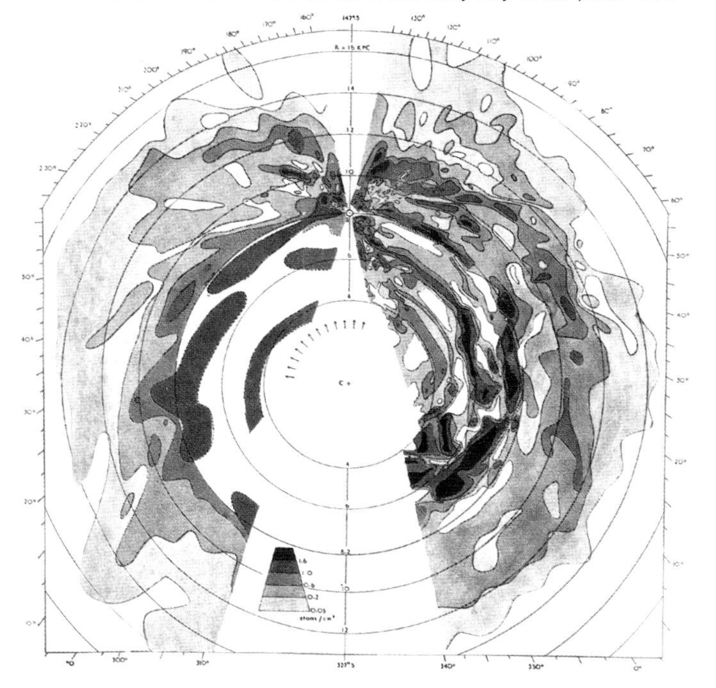

The gas clouds, and therefore the stars, furthest from the center are moving far too fast for the gravity of all the stars interior to their orbits to keep them bound to the Milky Way. In other words, the distant regions and stars in the Milky Way should have flown away from our galaxy billions of years ago. Something that is not in the form of stars and gas yet generates its own gravity seems to be present to keep the Milky Way from literally flying apart. This material is called dark matter. The general shape of this material is that the Milky Way is embedded in a massive sphere of dark matter such that almost ten times more dark matter exists than the matter in the visible stars and gas in the Milky Way. This same "hyper-rotation" effect has been detected in many other galaxies, so our Milky Way is not the only one with such a massive dark matter halo.

During the 1970s, astronomers also detected radio emissions from simple molecules such as carbon monoxide and formaldehyde. These molecules were being formed deep inside the hydrogen clouds in regions where gas densities exceeded 1,000 molecules/cm^3. When this molecular cloud data was added to the hydrogen maps still more details about the Milky Way's shape emerged. As many as six prominent spiral arms could be traced almost completely around our galaxy. Our sun and solar system is located between the more distant Perseus Arm and the Carina-

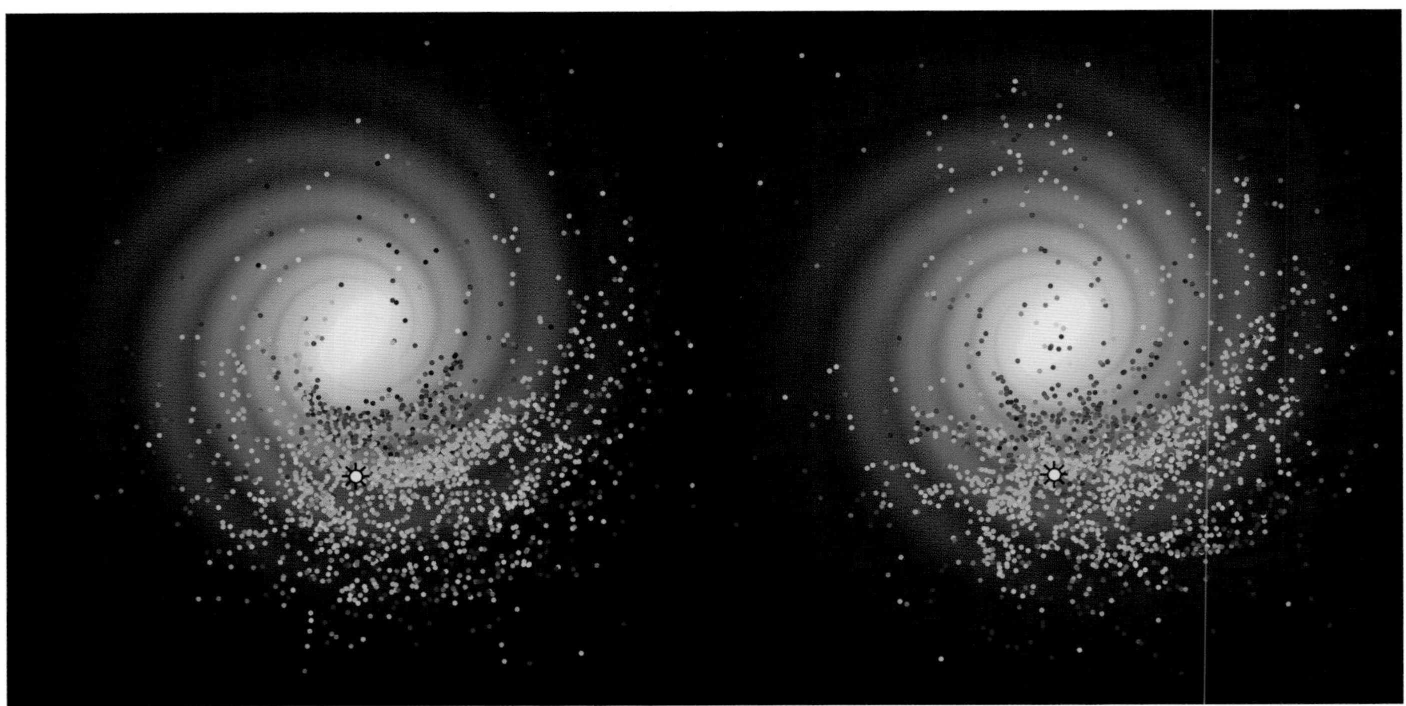

Above *Using a class of variable stars called Cepheid Variables, astronomers can plot the oldest of these stars (red), which are 400 million years old, and the youngest (blue), which are 30 million years old. Together, they help to map out the nearby spiral arms of our Milky Way. The sun is located near the yellow dot.*

Sagittarius Arm, on a little armlet of stars and clouds called the Orion Spur. At a distance of 26,000 light-years toward the constellation Sagittarius we encounter the vast bulge of stars in the nucleus of the Milky Way, and many signs that this region of space has been very disturbed over the last tens of millions of years. Gas motions seem no longer to follow the smooth rotations of the rest of the spiral disk.

After decades of research using many different kinds of tracers, from pulsars and planetary nebulae to galactic star clusters, we at last have a pretty good idea of what our Milky Way looks like, at least insofar as we can deduce its shape from inside measurements. That in itself is a remarkable achievement. It is almost akin to figuring out the street plan of Paris while standing at the base of the Eiffel Tower. There have been many attempts to render a realistic-looking Milky Way image from all of the astronomical data. The most intriguing and artful rendition was created in 2013 and combines data from a variety of astronomical sources together with details morphed from actual spiral galaxies to get the textures of the arm as accurate as possible, at least visually.

Although this is the simulated shape of the Milky Way with some artistic and "Photoshopped" elements added, can we find any examples of actual spiral galaxies that might look like our own? Fortunately, spiral galaxies are very common in the universe so all we have to do is match up our data with the shapes we see and find the closest matches.

Below *An artist's conception of the Milky Way galaxy.*

THE MILKY WAY'S CLOSEST TWINS

Here is a collection of spiral galaxies that each have elements shared by our Milky Way. An artistic rendering of what our galaxy looks like is unable to capture all of the subtle details of actual galaxies. Nature has a far-better gallery of models to choose from and in these we may catch a glimpse of our Milky Way's true appearance. There are some basic things we know to guide us. The Milky Way is a barred-spiral galaxy with very thick "flocculent" spiral arms. It has a diameter of about 100,000 light-years and a total mass, including dark matter, of about 1.5 trillion times our sun's mass. We also know the approximate sizes and distributions of its major spiral arms, as well as the patina of star-forming clouds and nebulae in the solar neighborhood. Here are a few of the possible "twins" to the Milky Way.

Above *UGC 12158 located 384 million light-years from the sun. Diameter: 140,000 light-years.*

UGC 12158

The closest match to the Milky Way in terms of diameter and mass is probably UGC 12158, located 384 million light-years from the Milky Way in the constellation Pegasus. It has a diameter of 140,000 light-years; about 40 percent larger than the Milky Way. It is a barred-spiral galaxy with a central bar that is somewhat larger than for our Milky Way. It also has about the same number of spiral arms, but they seem to not have nearly as much star-forming activity in bright nebulae as the spiral arms nearest our sun.

Above *NGC 3344, located about 20 million light-years from the sun. Diameter 40,000 light-years.*

NGC 3344

Located about 20 million light-years from the sun. It is only 40,000 light-years in diameter making it only 40 per cent as large as our Milky Way. It also has a rather large central bar about twice as large as we think is appropriate for our Milky Way, but the number of bright nebulae and star-forming activity seems a better match to our galaxy.

NGC 6744

A barred-spiral, located 30 million light-years from the sun, it is a massive galaxy some 200,000 light-years across, making it twice the diameter of our Milky Way. The size of its barred nucleus in proportion to the galaxy is a close match for our Milky Way, and the star-forming activity and nebulae are about as abundant as in our galaxy. But the arms are less distinct and there are numerous arm fragments and interconnections that radio studies of the Milky Way do not seem to show.

Above *NGC 6744, located 30 million light-years from the sun. Diameter 200,000 light-years.*

// The Milky Way as a cannibal

Like many other galaxies across the universe, our Milky Way does not travel alone through space. As one of the two largest galaxies in our Local Group of 54 galaxies, its gravity is able to reach out through intergalactic space and affect the motions of other nearby galaxies. Most of these, like the Large and Small Magellanic Clouds, are actually satellites to the Milky Way. They are close enough that their shapes are distorted by the Milky Way's enormous gravity. They can probably continue in this way for billions of years, but other dwarf galaxies near the Milky Way have not been so fortunate.

Detailed studies of the stars and gas within our Milky Way and in its immediate surroundings reveal that it has actually consumed several dwarf galaxies over the last ten

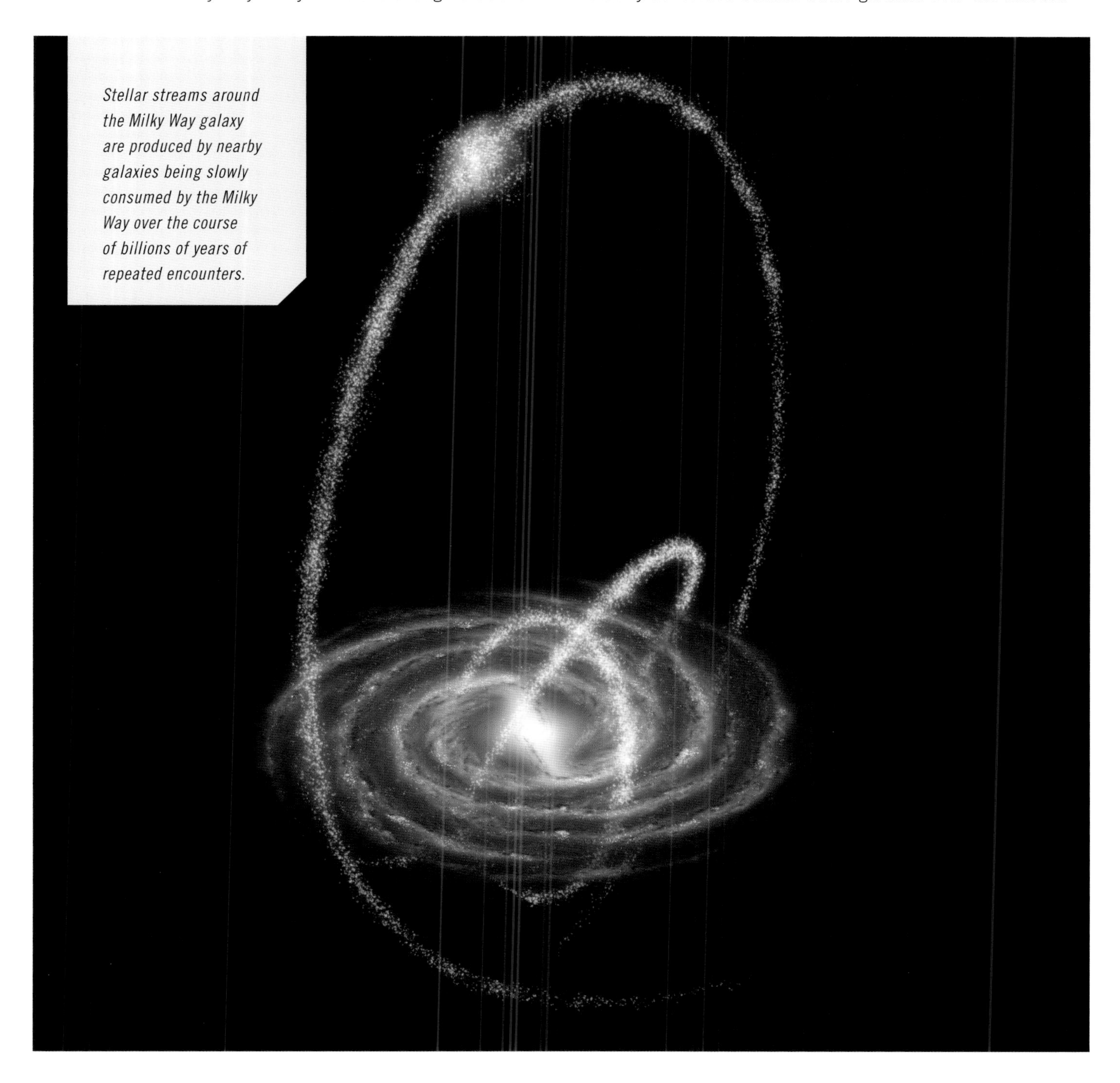

Stellar streams around the Milky Way galaxy are produced by nearby galaxies being slowly consumed by the Milky Way over the course of billions of years of repeated encounters.

Above *The merger with the Gaia-Enceladus galaxy occurred when our own Milky Way was still in its own infant state and did not display its modern spiral form. It is likely that its current spiral form was created at least in part by this encounter.*

billion years. The stars from these galaxies can still be found moving on their original orbits and form star streams and rivers of stars across the sky. One of these ancient meals was a galaxy now named Gaia-Enceladus. It was a large galaxy with about 50 billion stars, together with a retinue of eight of its own globular clusters. The Gaia satellite launched in 2013 made precise measurements of one billion stars near our sun and discovered some of the stars from this ancient collision. The globular cluster NGC 2808 is one of the largest orbiting our Milky Way and some astronomers speculate that this is actually the core of this ancient galaxy. More recently, the Sagittarius Dwarf galaxy has made several orbits through the disk of the Milky Way with the next event to occur in another 100 million years. Each time it loses more of its mass until eventually it will fully merge with our galaxy and vanish. But the worst is yet to come.

In another 4 billion years, the nearby Andromeda galaxy will collide with the Milky Way traveling at a speed of about 435,000 mph (700,000 km/h). During the ensuing billion years, the pieces of these two giant galaxies will thrash about before settling down into a new galaxy some have called Milkomeda. This new galaxy will resemble a giant elliptical galaxy surrounded by star-forming clouds of gas and dust. The central supermassive black holes in each of these two galaxies will, in time, merge together and produce a quasar-like core of matter until most or all of the remaining interstellar matter is consumed. The black hole will then lay dormant perhaps for billions of years since there are no similar, large galaxies known to be candidates for collisions in at least the next ten billion years to come.

Below *This illustration shows a stage in the predicted merger between our Milky Way galaxy and the neighboring Andromeda galaxy, as it will unfold over the next several billion years. In this image, representing Earth's night sky in 3.75 billion years, Andromeda (left) fills the field of view and begins to distort the Milky Way through its gravitational "tidal" pull.*

PLANETARY SYSTEMS

Beginning in the 1990s, the advent of powerful supercomputers and telescopes allowed astronomers to uncover the details of how planetary systems form. Not only could detailed theories be put to the test with supercomputer models, but spectacular telescopic images of solar systems in the making became an increasingly common resource for uncovering the major stages in planet formation. The discovery of thousands of planets orbiting distant stars also provided a renewed interest in discovering habitable worlds beyond our solar system. No longer a figment of science fiction, exoplanet research has blossomed in the last decade, revealing just how unique our own solar system appears to be.

Planetary systems abound in the Milky Way and may easily outnumber the population of stars themselves. Although scientific knowledge about their specific details are still rudimentary, enough is known for artists to render them in plausible detail. Some rogue planets may not even be gravitationally tethered to stars, but instead are free to roam interstellar space.

// Protoplanetary disks

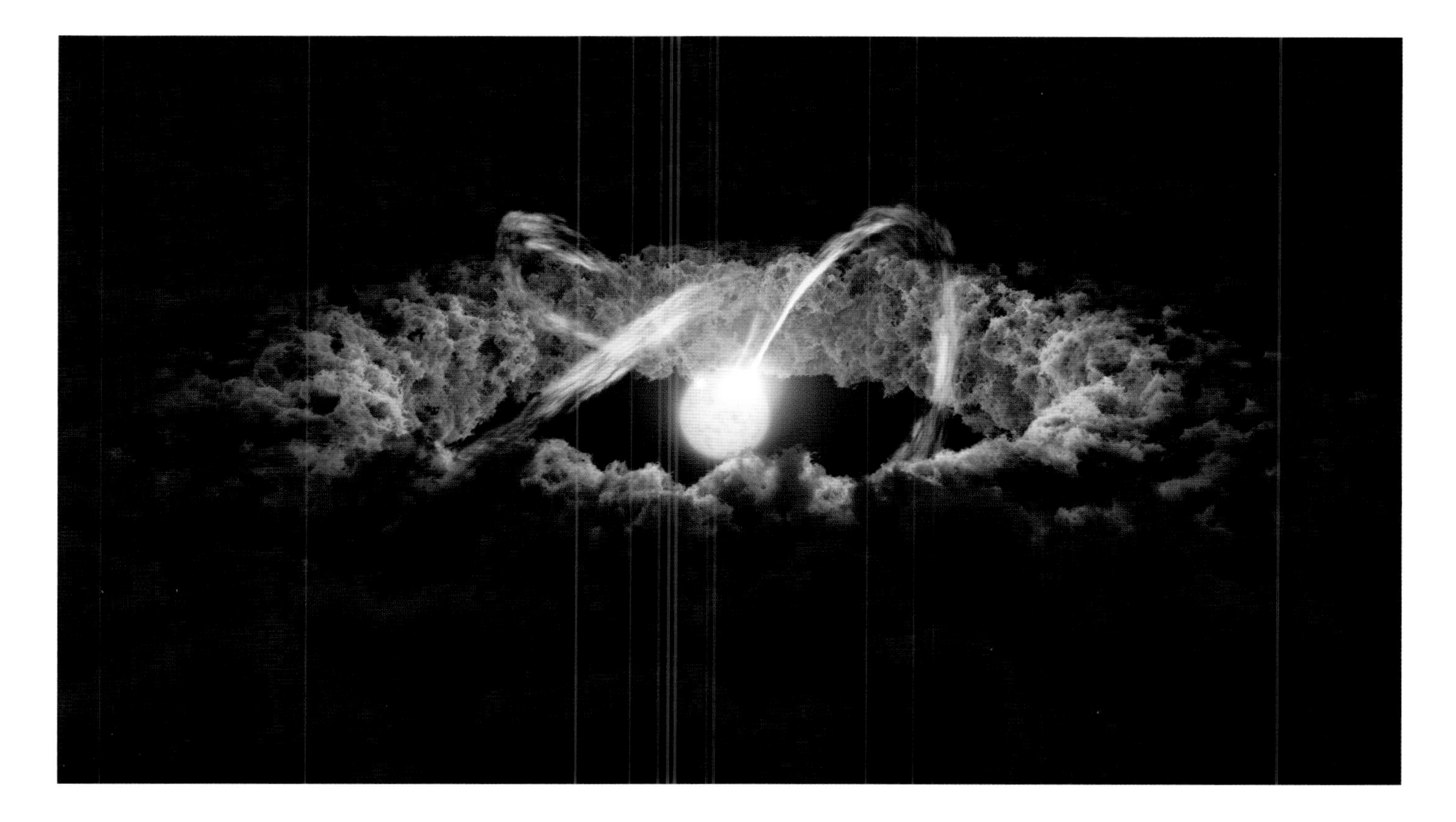

Above *An artistic conception of a protoplanetary disk.*

Since Immanuel Kant proposed the elegant idea in 1775 that planets form from rotating disks of gas, astronomers have found numerous examples of these swirling discs surrounding infant stars. Meanwhile, astrophysicists have developed sophisticated mathematical models in supercomputer simulations of how such gas should behave and how planets could form from them. Here's how it works.

Following the collapse of a portion of a dense interstellar cloud into a proto-stellar globule, the center of this globule continues to grow in density and will eventually accumulate enough mass to become an infant protostar. Meanwhile, because the gas cloud is rotating, it forms a flattened disk around the protostar. The protostar heats the gas in the disk to thousands of degrees at its closest to the protostar, and a mere few hundred degrees above absolute zero at its outermost extremity. Within this temperature range, complex compounds can form. Silicate-rich minerals with high melting points become common in the inner disk regions, and ices of water and other volatile gases become common in the frigid outskirts of the "protoplanetary" disk. The material in this disk starts out to be rich in dust grains only a few microns in size, but through collisions in the dense material of the disk, they grow quickly to millimeter, centimeter, and meter-sizes. These collisions have to be gentle enough that they do not shatter the growing mass. Over millions of years, the disk becomes enriched with innumerable bodies spanning a wide range of sizes and masses, from sand grain-sized objects up to planetoids hundreds of miles across.

When planets form, they accumulate from materials common to the regions where they are located. In the inner disc zone, which can extend as distant as the orbit of the planet Mars in our own solar system, rocky planets rich in silicate and iron compounds are common, while outside this zone, objects rich in ices with trace quantities of silicates become common. The wild card in this process occurs when an object accretes enough mass to grow larger than about ten times the mass of Earth. At this point, it acts like a cosmic vacuum cleaner with its huge and growing gravitational influence. It grows rapidly within a few million years to become a giant planet like our own Jupiter, Saturn, Uranus, or Neptune. The resulting swept-out zones can be so large they can actually be seen in the protoplanetary disk itself. One such example is the star HL Tauri located only

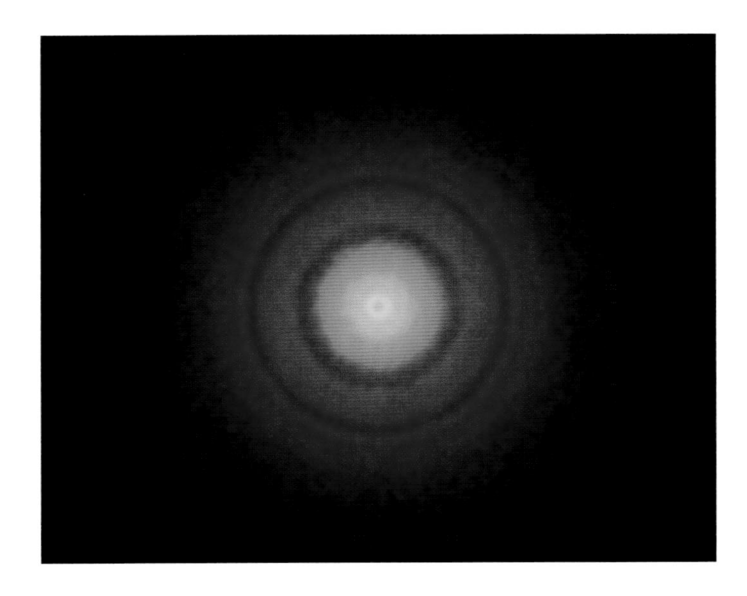

Left *This image of the star TW Hydrae was taken with the ALMA telescope and reveals the gaps in the protoplanetary disk caused by the formation of several planets. The star is 175 light years from the sun and is only 10 million years old. The outer gap is located 22 AU from the star; a distance comparable to the orbit of Uranus.*

450 light-years from our sun. Smaller planets take longer to form because they do not have a huge gravity-assist to accumulate matter, so that direct collisions at very slow speeds are necessary. If the collision speeds are too fast, the forming planet shatters after impact and the process of planet building has to start nearly from the beginning. Our Earth probably required 20 million years to form. Even today, it is still accumulating more than 56,000 tons (50,802 tonnes) of asteroidal and meteoritic material every year, so technically it is still a forming planet.

The search for actual planets orbiting nearby stars has been a decades-long hunt made possible by dramatic advances in telescope design and imaging technology. Among the first planets beyond our solar system to be directly imaged were those orbiting the star HR8799, located 120 light-years from our sun. Its three Jupiter-sized planets were imaged by the Palomar Observatory's 200-inch (508-centimeter) Hale telescope, improved by the addition of new technology such as a stellar coronagraph. This device eclipses the bright light from the star itself so that the million-time-fainter light from orbiting planets could be detected. As technology continued to improve, astronomers have been able to actually spot the formation of massive planets within their protoplanetary disks in young stars such as AB Aurigae, PDS 70b, and TW Hydrae.

Below *An artist's impression of a planet orbiting the star HR8799.*

// Planets outside our solar system

Beginning in the mid-1990s, astronomers began searching for signs of planets orbiting nearby stars using advanced spectroscopic instruments and devices that measure very small changes in a star's brightness. As a planet orbits a star, its gravity tugs at the star causing it to move. This motion can be detected to an accuracy of a few meters per second using spectrometers. Sensitive light meters search for periodic dips in the brightness of a star as its orbiting planet eclipses the star, dimming its starlight. However, the very first confirmed planets were not found by these transit-eclipse and spectroscopic methods.

In 1990, Polish astronomer Aleksander Wolszczan conducted detailed studies of a pulsar he had discovered called PSR B1257+12 located 2,300 light-years from the sun and noticed something odd about its orbit and spin. Two years later he and American astronomer Dale Frail concluded that the period anomalies were caused by three small bodies orbiting the pulsar within only 43.5 million miles (70 million kilometers), and that these objects would be less than twice the mass of our moon. Then, in 1995, a Jupiter-sized planet was detected orbiting the star 51 Pegasi only 50 light-years from our sun by astronomers Michel Mayor and Didier Queloz at the University of Geneva using spectroscopic techniques. It was the first "exoplanet" found orbiting an ordinary sun-like star. Between 1995 and 2011, some 67 additional exoplanets were discovered orbiting other stars using a combination of transit and spectroscopic techniques.

In 2008, NASA launched a spacecraft called Kepler, which imaged nearly 160,000 stars in the direction of the constellation Cygnus. Every 29 minutes, Kepler recorded and averaged the brightness of these stars and the data was transmitted back to Earth. Astronomers then used computer programs to search for signs of periodic starlight changes that could be a planet passing across the face of the star as viewed from Earth. Its first five exoplanet transits were reported in 2011, but by the end of its mission in 2018 it had discovered more than 4,800 exoplanets, of which 2,662 have been independently confirmed. In just over 20 years, we went from our solar system containing the only known planets in the universe, to a bewildering catalog

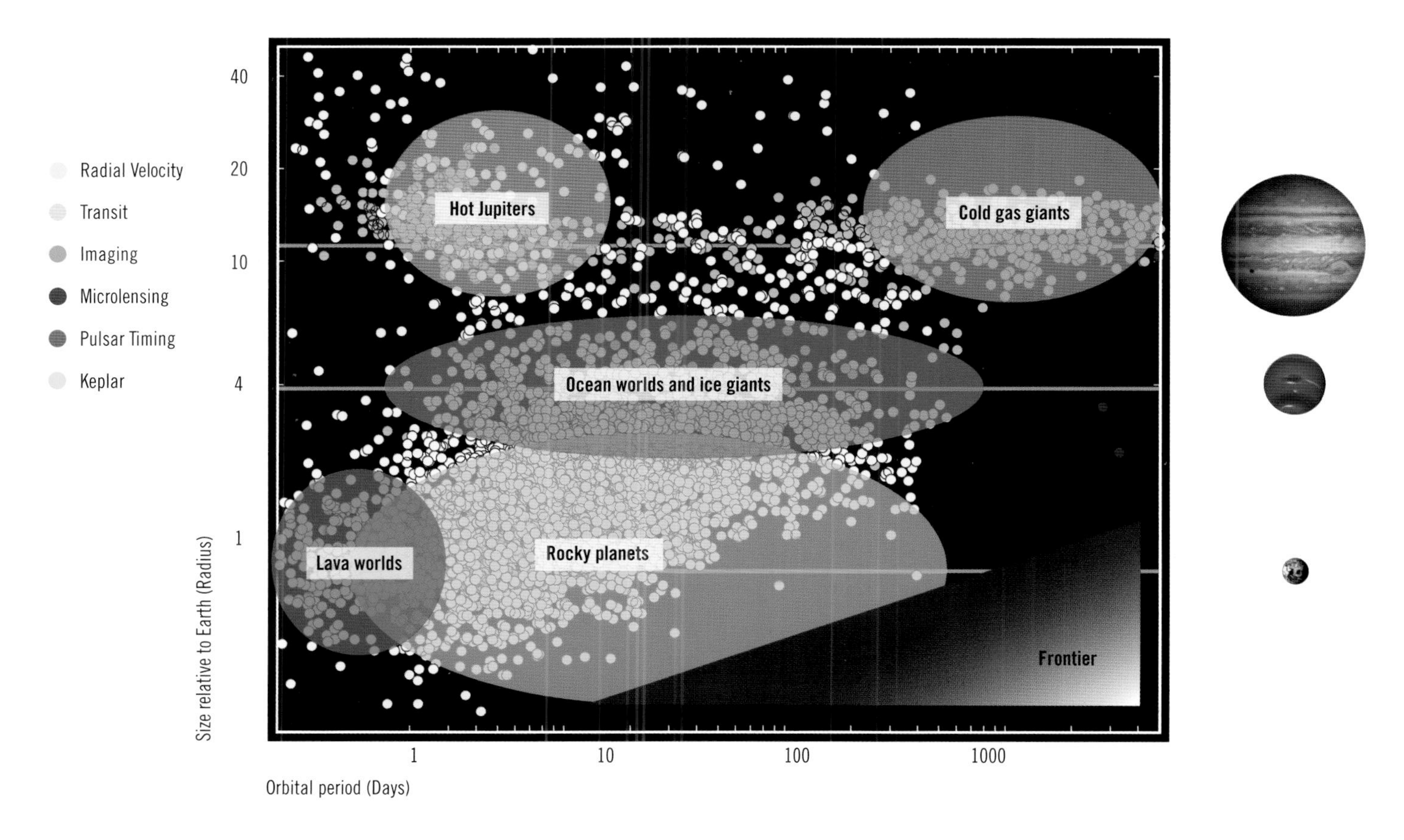

Below *The exoplanets detected by the Kepler mission can be grouped into specific types. The horizontal lines mark the sizes of Jupiter, Neptune, and Earth. The shaded gray triangle at the lower right marks the exoplanet frontier that will be explored by future exoplanet surveys. These are very small planets located at great distances from their stars.*

A table of nearby exoplanets

Name	Distance (Lys)	Star Type	Orbit Distance (AU)	Orbit Period (days)	Mass (Earth=1)	Exoplanet Type
Proxima Centauri b, c	4.2	Red dwarf	0.05, 1.5	11, 1,928	>1.2, 7	Earth, Super Earth
Barnard's Star b	5.9	Red dwarf	0.43	232	>4.2	Super Earth
Wolf 359 b, c	7.9	Red dwarf	1.8, 0.02	2,940, 2.7	>44, >3.8	Super Earths
Lalande 21185 b	8.3	Red dwarf	0.07	9.8	>2.9	Super Earth
Epsilon Eridani	10.4	K-Main Seq	3.4	2,500	450	Jovian
Lacaille 9352 b, c	10.7	Red dwarf	0.07, 0.12	9.3, 21.8	>4.2, >7.6	Super Earths
Ross 128 b	11.0	Red dwarf	0.05	9.9	>1/4	Earth

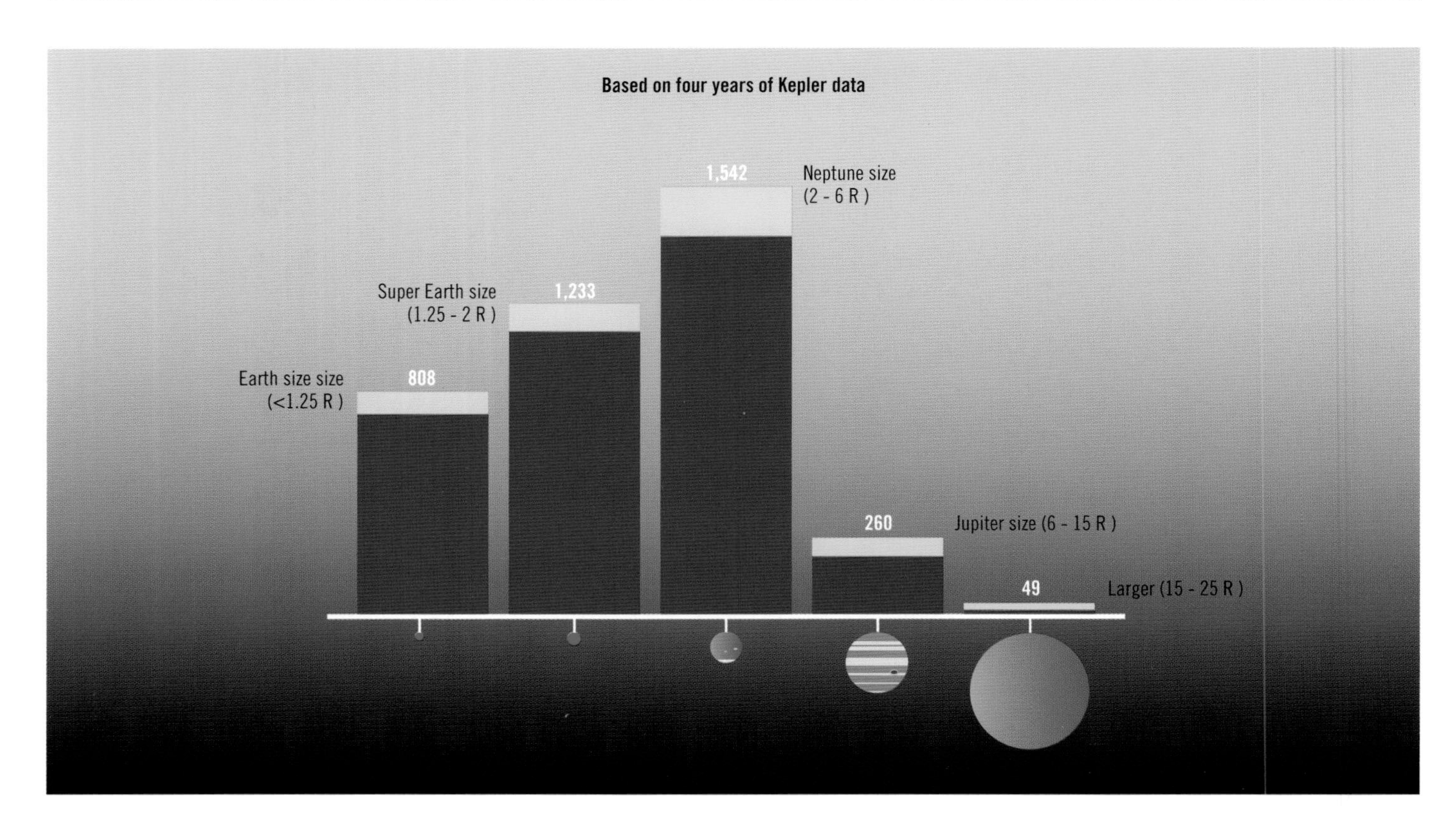

Above *Among the thousands of exoplanets detected to date, those the size of our Earth are not uncommon and represent about one quarter of the known exoplanets.*

of diverse exoplanets of every size and composition. For the first time in history, the existence of planets beyond our solar system became a fact of life and not merely a science fiction staple. The way that astronomers use this data to "see" what these exoplanets might look like is a fascinating study in ingenuity and the application of basic physics. Astronomers can use the period of the transit together with a knowledge of the mass of its parent star to determine the distance of the planet from its star. From the amount of light dimming, the diameter of the planet can be calculated. Also, with data from spectroscopic studies the gravitational dance between star and planet revealed by minute speed changes in their orbits could be quantified and from this the mass of the planet could be found. Once its distance from its star is known, you can calculate the planet's surface temperature, and from its mass and diameter you can determine its density. Within a short time, astronomers had discovered thousands of planets and determined the sizes and densities for many of them. What they discovered was that planets come in various size ranges and their properties based on density could be compared with planets in our own solar system.

With current technology it is easier to detect the exoplanets that are the size of Neptune or larger so it is no wonder that from the large population of exoplanets we see more of these planets than the smaller Earth-sized ones.

More than 96 percent of all detections are for exoplanets more than twice the size of Earth. But the detection of Earth-sized exoplanets steadily increases every year. It is now estimated that every star has at least one exoplanet, and that one in five stars has one that is Earth-sized. Because of the great variety in sizes, masses, and surface temperatures an almost unimaginable range of possibilities exist for what these worlds may look like, ranging from ice-bound planets and planets completely covered with an ocean, to literal hell-holes where the sky rains liquid iron.

An exoplanet's size, mass, and density determine quite a bit about what kind of exoplanet it will be. Exoplanets the size of Earth or smaller are rocky bodies like the inner planets of our solar system. At a mass of about five times Earth, these Neptune-sized objects will have a thick watery ocean and likely no continental land masses. By the time you reach a size about ten times Earth, the object is a giant ball of gas like Jupiter or Saturn. The formation energy of these exoplanets is high enough that they continue to be bright infrared sources. There is,

however, a limit to how large an exoplanet can get. Once it reaches a mass of about 13 times that of Jupiter, its own gravity compresses the exoplanet's core, producing a high-enough temperature to trigger thermonuclear fusion using deuterium. Instead of an exoplanet, we now have a brown dwarf star. Fully-fledged stars burn hydrogen into helium, and this critical mass is about 75 times that of Jupiter for what are called red dwarf stars. The properties of an exoplanet also depend on the kind of star it is orbiting and its distance from that star.

Above Left *With a surface temperature above 3,632°F (2,000°C), the exoplanet Wasp-76b located 640 light-years from Earth offers the visitor clouds of iron vapor that condense into liquid-iron rain on blustery days when wind speeds exceed 3,100 mph (5,000 km/h).*

Above *This artist's illustration allows us to imagine what it would be like to stand on the surface of the exoplanet TRAPPIST-1f, located in the TRAPPIST-1 planetary system about 40 light-years from Earth in the constellation Aquarius.*

// Internal structure

A crude indicator of the internal structure of a planet is simply its density. Planets rich in iron and nickel have densities in excess of 5 g/cc such as Mercury (5.4 g/cc). Planets rich in silicate compounds like our Earth's crust have densities near 3 g/cc. If the planet contains significant amounts of ices and water it will have densities closer to 2 g/cc, such as Neptune (1.7 g/cc). Finally, if the planet is mostly a ball of gas it can have densities as low as 1 g/cc, such as Jupiter (1.33 g/cc) or Saturn (0.7 g/cc). This ball-park estimate for its structure can be estimated just by knowing the planet's mass and its diameter. But astronomers have detailed physical theories for how matter accumulates into a sphere, and how its internal layers will change depending on its internal temperature, pressure, and composition. The composition can be guessed by knowing how far from its star the planet formed. This distance also determines how hot its atmosphere and surface will be from the absorbed sunlight. For example, very small planets close

Below *Exoplanets comparable to Earth in size and mass are likely to have a similar interior structure, although the proportions will vary. Super-Earths more than five times the mass of Earth may have surfaces covered by deep oceans, while the gravitational fields of still-larger exoplanets will attract dense gaseous atmospheres like those of Jupiter and Saturn.*

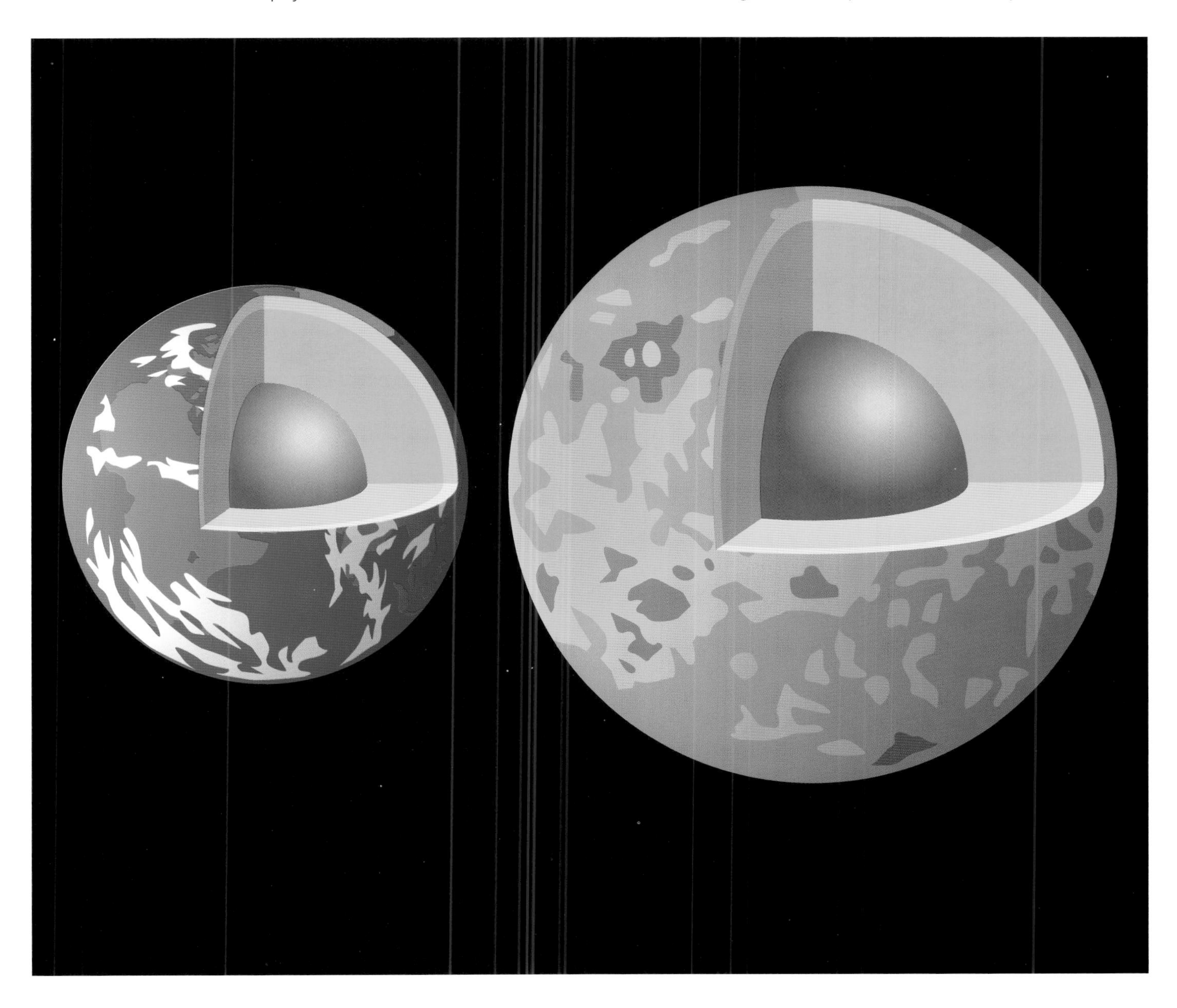

Above *A simulated comparison of sunsets on Earth and various exoplanets.*

to their stars will not have thick atmospheres because their gravities are insufficient to keep their atmospheres from evaporating into space, which is the case for our planet Mercury. However, very massive planets like Jupiter can get so close to their stars that their surface temperatures are over 2,372°F (1,300°C) without significant atmosphere loss. These are called Hot Jupiters and there are known to be over 300 of these exotic exoplanets. In the case of 51 Pegasi b, called Bellerophon, it is actually evaporating and sports a comet-like tail as it orbits its star every four days.

Objects similar to our Earth in size can have complicated interiors. If they orbit far enough from their stars so that they are not too hot, the more massive "super-Earth" exoplanets can have a surface covered by a thick ocean and a dense atmosphere. They can have thick crusts that prevent continental drift and reduced volcanic activity. Exoplanets that are in Earth's mass and size ranges can have limited planetary oceans and some continental plate tectonics. For exoplanets significantly smaller than Earth such as Mars-like worlds (one-tenth the size of Earth) they will generally have limited to non-existent plate tectonic activity because their interiors are cooling off too rapidly. This can also mean that they are unable to maintain a strong planetary magnetic field to protect their atmospheres against evaporation through interaction with their solar winds.

When Earth-sized or larger exoplanets are found far from their stars where temperatures do not permit liquid water, these bodies develop very thick mantles of various ices whose properties vary in their interiors. They can have thick crusts of solid ice floating atop deep planetary oceans that, in turn, are situated on top of a rocky Earth-like core rich in silicates. So, astronomers can create estimates for what exoplanets might look like given their sizes, densities, and distances from their stars. When combined with a bit of technical artistry, one can even imagine what their surfaces might look like were one to stand, or float, on them.

// Habitable zones

Exoplanets that orbit close to their stars can reach very high surface temperatures of more than 1,832°F (1,000°C). For some exoplanets such as 55 Cancri e, also called the Hellfire Planet, its oceans consist of molten lava and its "rain" consists of droplets of molten rock. Another planet, HD 209458b is so close to its star that it is actually evaporating like a comet shedding a huge tail.

These are not the kinds of worlds of interest to us because liquid water, an essential ingredient for life, could not exist on their surfaces. By knowing the type of star and the exoplanet orbit distance, it is possible to mathematically calculate where a planet would need to be so that liquid water could exist on its exposed surface assuming it had an atmosphere that would prevent water escape by evaporation. This range of orbits can be calculated for every star and is called the habitable zone. For our solar system, it extends from about the orbit of Venus to the orbit of Mars. Inside this zone, surface temperatures favor water evaporation. Outside this zone favors water in the form of ice.

Another important factor is the composition of the exoplanet atmosphere. If it contains more than about 1 percent carbon dioxide, greenhouse heating will cause the surface temperatures to grow rapidly, releasing more greenhouse gases such as carbon dioxide and water into the atmosphere. Without surface oceans and plate tectonics, the carbon dioxide abundance will continue to increase

Below *This illustration shows the dozens of exoplanets discovered by Kepler that are less than twice the size of Earth and orbit in or near their stars' habitable zones. These planets are likely to be rocky or ocean worlds and are the most likely candidates for supporting life.*

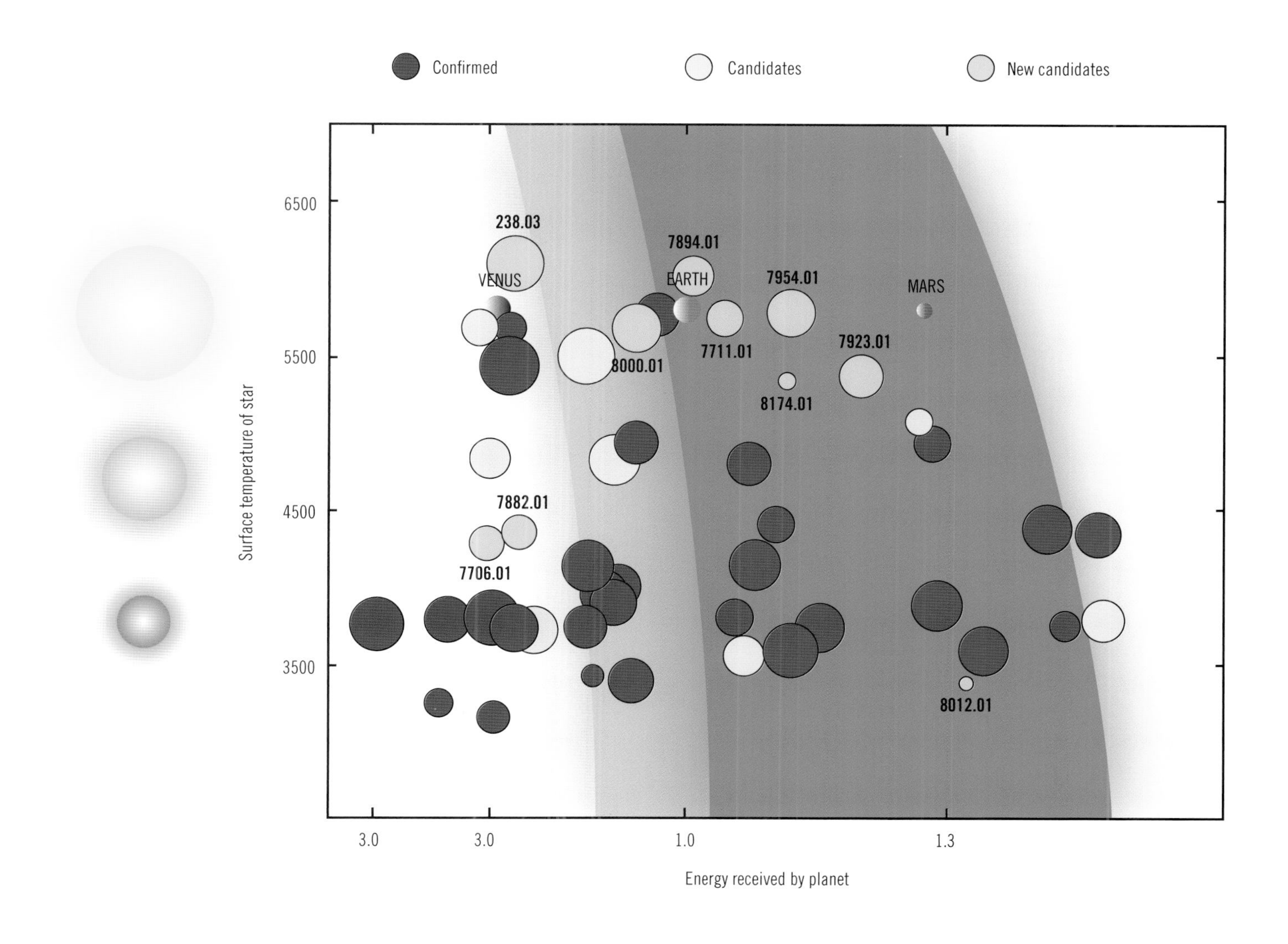

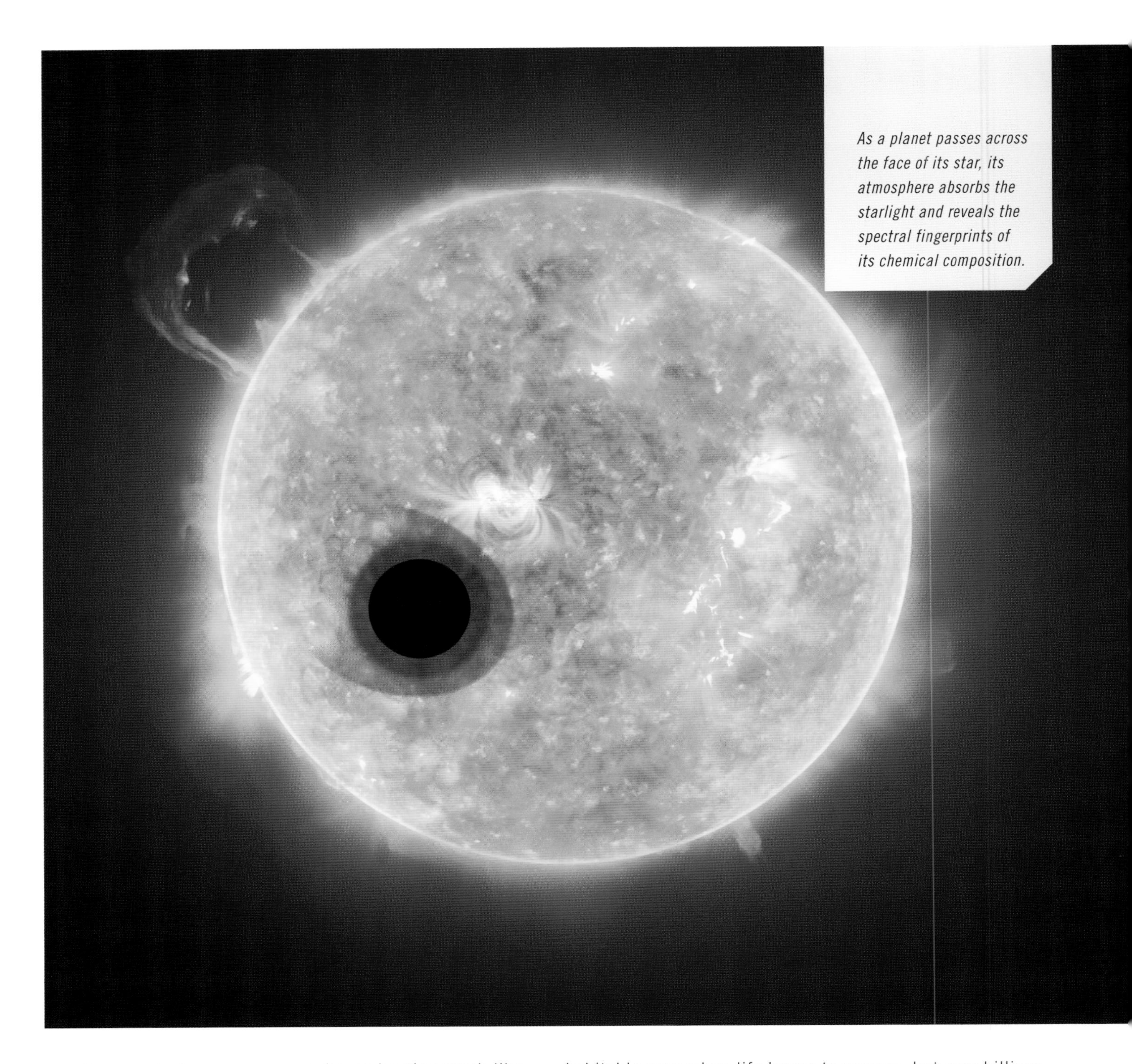

As a planet passes across the face of its star, its atmosphere absorbs the starlight and reveals the spectral fingerprints of its chemical composition.

over time producing runaway greenhouse heating much like Venus in our solar system. Astronomers can tell whether a planet is in the habitable zone of its star, but cannot yet measure the atmospheres to see whether too much greenhouse gas makes the exoplanet a Venus-like world even in the habitable zone.

Another feature of habitable zones is that they do not remain fixed in space and time. As a star evolves, it becomes more luminous and so its habitable zone moves further from its star. This has the troubling consequence that an exoplanet may have started out in its star's habitable zone when life began to emerge, but over billions of years as its star evolves, the exoplanet may now find itself outside its star's current habitable zone, which could cause the extinction of life on that world. Our own Earth, for example, is close to the inner edge of the sun's current habitable zone, but in just over one billion years Earth will find itself inside this zone and so life on this world will become extinct. It is a staggering thought to dwell upon how many worlds in the Milky Way have lost their entire biospheres in the last few billion years due to the inexorable migrations of their habitable zones.

// Earth analogs

Above *Proxima Centauri is a red dwarf star with frequent flares. If its orbiting planet has an atmosphere, and a magnetic field, it is likely that spectacular aurora will be seen, but their colors will depend on whether oxygen, hydrogen, or nitrogen are present.*

Since the 1990s, when exoplanets were first detected, astronomers and the general public have been fascinated by the prospects of finding a twin to our Earth: an exoplanet of nearly identical size and mass orbiting within the habitable zone of its star. We still, as yet, cannot measure their atmospheres to see if they are greenhouse hells or have Earth-like chemistries rich in nitrogen, but they offer the start of a growing list of exoplanets that will merit future scrutiny. Across the entire Milky Way galaxy, it is estimated

A table of nearby Earth-like planets in their habitable zones

Name	Distance (Lys)	Star Type	Orbit Distance (AU)	Orbit Period (million km)	Mass (Earth=1)	Temperature (K)*
Proxima Centauri b	4.2	M-dwarf	7.5	11.1	>1.3	234
Luyten b	12.4	M-dwarf	13.5	18.7	2.6 – 3.2	206 to 293
Tau Ceti e	12	G-type	82.5	163	>3.9	285
Kapteyn b	13	M-dwarf	25.2	48.6	>4.8	205
Wolf 1061c	13.8	M-dwarf	5.7	17.9	>4.34	275
Gliese 667 Cc	23.6	M-dwarf	18.8	28.1	>3.8	277
Trappist-1d	39	M-dwarf	3.3	4	0.30	258
Trappist-1e	39	M-dwarf	4.4	6	0.77	230
Trappist-1f	39	M-dwarf	5.6	9	0.93	200
Trappist-1g	39	M-dwarf	6.9	12	1.15	182

*Earth=255

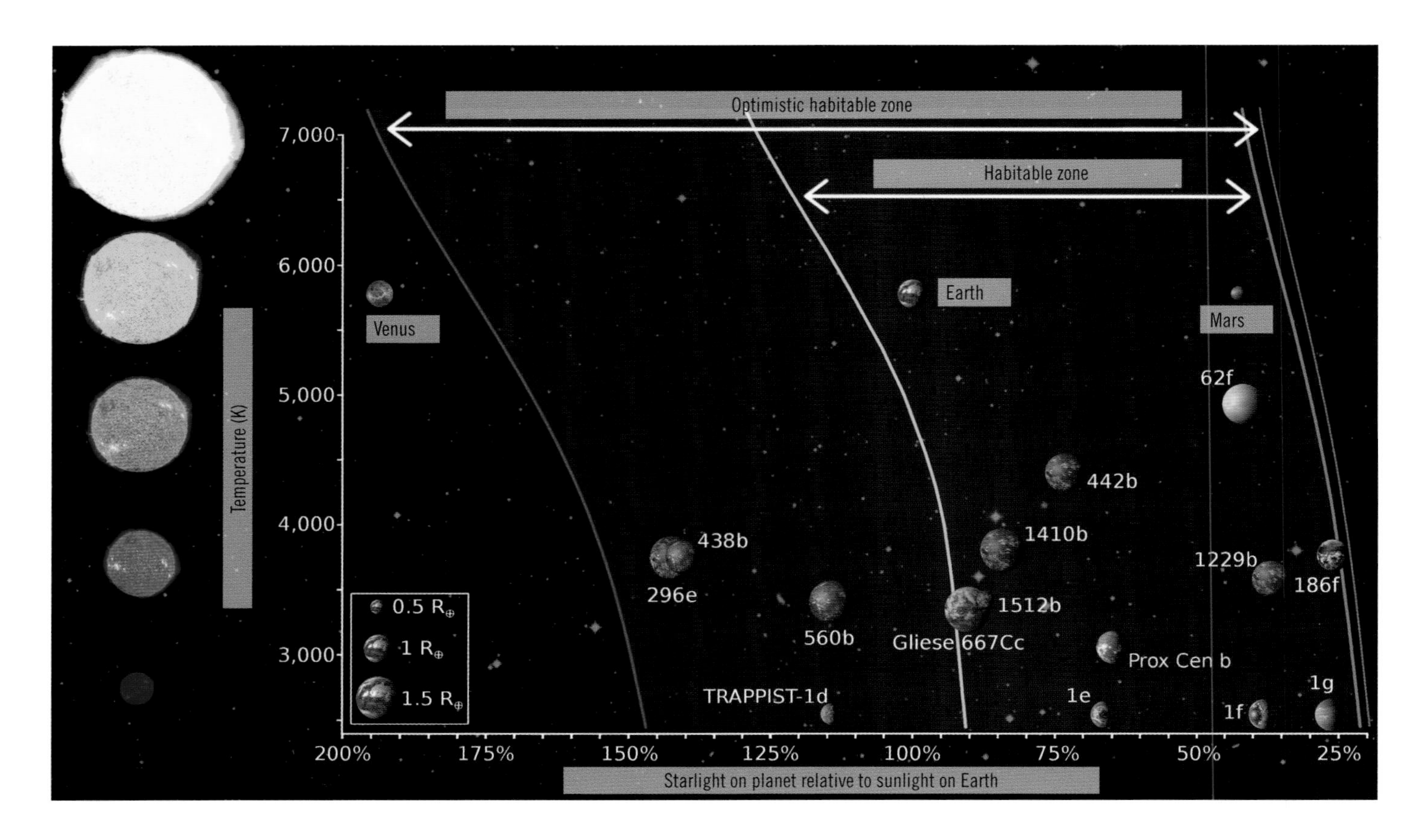

Above *This diagram shows the habitable zone for stars of different temperatures, as well as the location of terrestrial-size planetary candidates and confirmed Kepler planets described in new research. Some of the solar system terrestrial planets are also shown for comparison.*

that there could be as many as 40 billion Earth-analogue worlds, and many of these will be found orbiting, as our own Earth does, around a lovely yellow star. So far, several dozen of these Earth-analogs have already been detected by the Kepler observatory and other ground-based means.

The most intriguing of these are Proxima Centauri b, located 4.2 light-years from the sun; Kepler 1649c, located at a distance of 302 light-years; Kepler 186f, at 560 light-years; and Kepler 452b at a distance of 1,400 light-years. Other nearby stars also have detectable exoplanets but these are either too large or do not orbit in their habitable zones.

Science fiction authors have long been intrigued by our nearest star Alpha Centauri and its companion red dwarf star Proxima. Stories abound with tales of intrepid explorers taking rocket ship journeys lasting years at a time to discover whether planets exist there, and banking on being able to colonize them. Today, we can determine whether this is true by actually detecting these worlds and studying them long before any astronauts have to make the century-long journey. Astronomers continue to search for hints that there are in fact planets orbiting the Alpha Centauri stars A and B but have already been rewarded by detecting planets around Proxima Centauri.

Proxima is a red dwarf star about one-fifth as luminous as our sun. Its dull red light would bathe exposed planetary surfaces in a disturbing crimson glow. Proxima is also known for the flare-ups of particles and radiation that happen every few years, making any close-in planets a radiation hazard. Proxima has two known exoplanets: Proxima Centauri b is about 1.2 times Earth's size and orbits the star every 11 days at a distance of 4.7 million miles (7.5 million kilometers), which is well inside Mercury's orbit. Proxima Centauri c is a super-Earth exoplanet with a mass of seven times our Earth and orbits once every five years at a distance of 137 million miles (220 million kilometers), which is similar to Mars. Because of the dimness of Proxima Centauri, its habitable zone is pulled close to the star, and the Earth-like Proxima Centauri b is actually inside the habitable zone. Because of the flaring activity of the star, however, it is unlikely to be habitable unless life has developed below its crust or surface life is nocturnal or has developed a fondness for lead coatings and armor. To make matters worse, the intense winds from this active red dwarf star may also have completely stripped the planet of its atmosphere.

// The late bombardment era

The intensive studies of Kepler's exoplanets have also revealed a number of important discoveries about the formation and evolution of our own solar system. Several dozen exoplanet planetary systems have been discovered so far, but none of them look anything like our solar system. In some cases, such as GJ 876 or Upsilon Andromedae, the giant planets are crowded together in orbits closer than Mercury's orbit around our own sun. For other exoplanet systems, such as Trappist-1, there are no giant planets at all, but the small rocky worlds orbit closer to their star than Mercury. There doesn't seem to be any rhyme or reason to where the exoplanets are located or how large they are, unlike our solar system where the giant planets are the most distant and the rocky planets are closer to the sun. Astronomers have begun to understand why this might be so.

Planets are formed inside protoplanetary disks rich in gas, dust, asteroids, and other material. Although forming planets have a gravitational effect on the surrounding disk material, it is also true that the disk material affects the forming exoplanet. This causes a friction-like force that forces the orbits of the forming planets to slide closer to the central star. This process stops only after the disk material has dissipated. As a result, a giant planet may form far from its star but over millions of years slide closer and closer to its star. The number of Hot Jupiter exoplanets that have been discovered orbiting only a few million miles from their star suggests that this friction process could be so dramatic that in some cases the star actually eats some of its own exoplanets over time. Astronomers have suspected that this process occurred in our own solar system because the element abundances of Uranus and Neptune seem to be more appropriate for material further from the sun than where these planets are today. This process may also have occurred for Jupiter but its progress toward the sun was halted before it could invade the inner solar system and eject the inner planets, including the forming Earth.

The inward migration of massive planets causes any smaller planets to be ejected from the planetary system

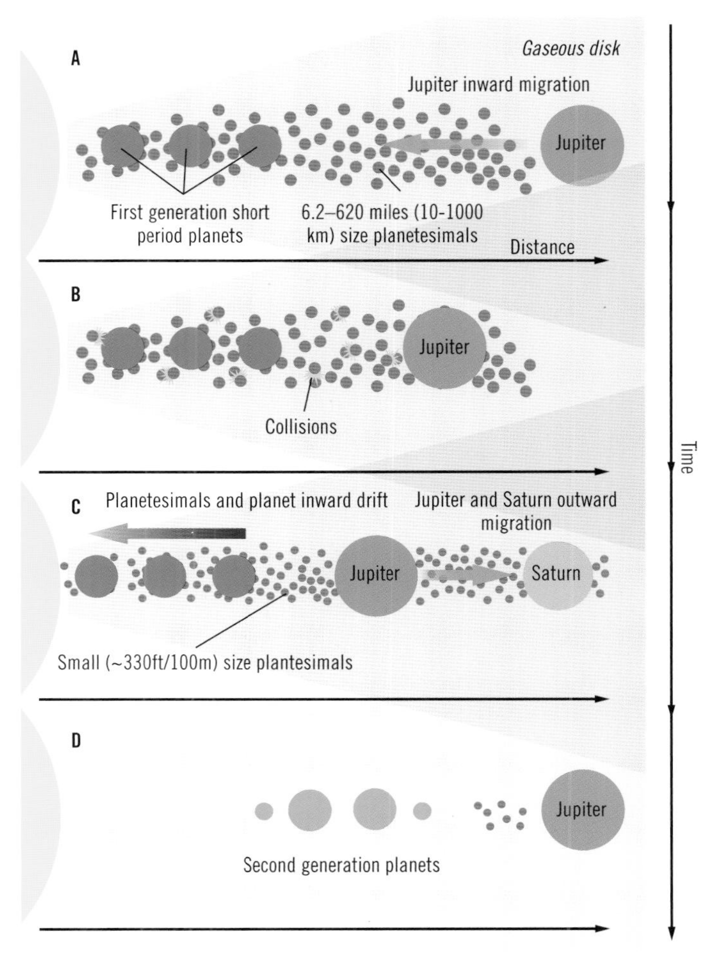

Above *Description of the early solar system evolution with massive, migrating giant planets. Disruption of asteroid and comet orbits led to the late bombardment era in our solar system's early history.*

Below *The exoplanet orbiting the star HD189733 once every two days is just one among many "Hot Jupiter" planets that formed far from their stars and drifted inwards.*

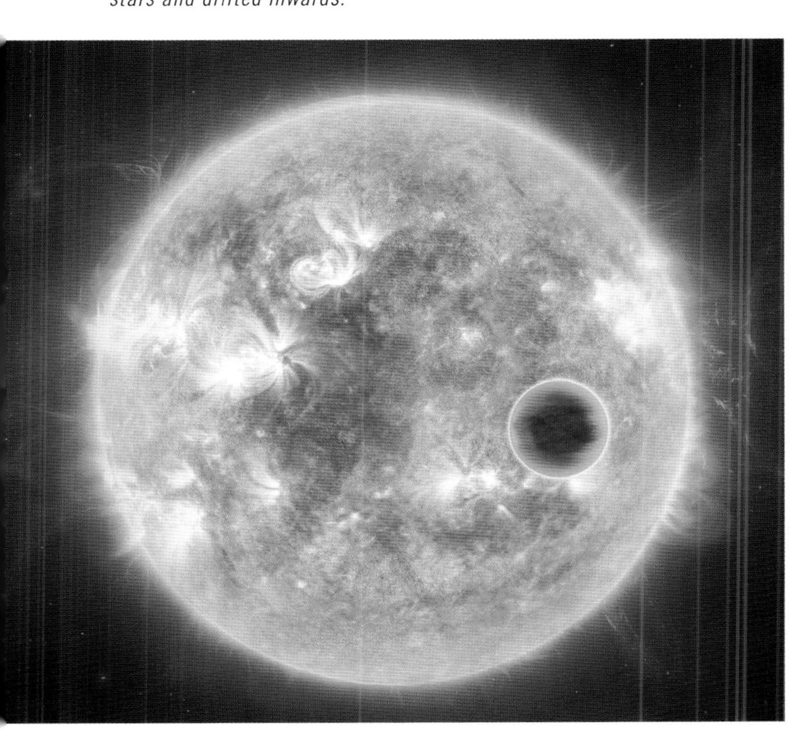

millions of years after the giant planets have reached their final orbits. Then there is a possibility for smaller planets to re-form from whatever stray disk material has remained. This leads to many lop-sided planetary systems with giant planets close to their star with smaller planets further away. With too many migrating giant planets, smaller planets may have no chance to form at all before the protoplanetary disk dissipates completely.

Below *An artistic rendering of the late bombardment era as viewed from Earth's surface more than four billion years ago. Planet migration events involving Jupiter and Saturn may have been responsible.*

OUR SOLAR SYSTEM

Modern astronomers have benefitted from a Golden Age in planetary exploration. Robotic probes and spacecraft have provided close-up views of planets' surfaces and environments. We have surveyed all of the major bodies in our solar system, dug into the soils of Mars in search of life, and even mapped the surface of the distant dwarf planet Pluto. Humans have set foot upon the moon, while at the same time we have drawn up plans for colonies on Mars. The bewildering landscape of our solar system reveals a complex and violent past, but also includes small enclaves where life might take hold in sub-surface oceans on Europa, and permafrost on Mars.

Although this artistic rendering is not drawn to scale, it gives a sense of the many ingredients that shape our solar system from the surface of the sun to the most remote asteroids and interplanetary gas clouds.

// The solar system as a whole

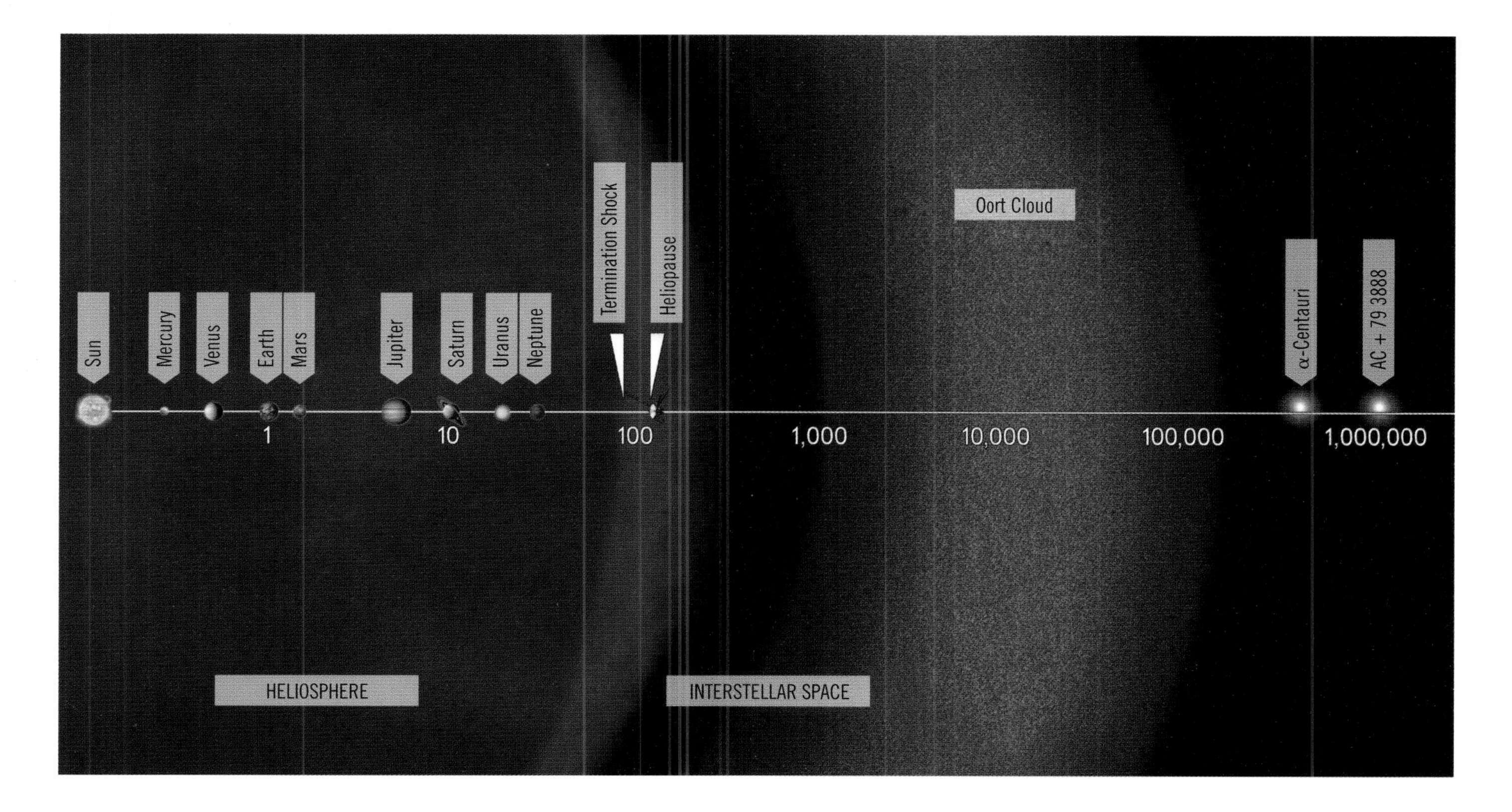

Above *This diagram puts solar system distances in perspective. The scale bar is in astronomical units, with each distance interval beyond 1 astronomical unit (AU) representing ten times the previous distance. One AU is the distance from the sun to Earth, which is about 93 million miles (150 million kilometers). The sizes of the sun and planets are not drawn to scale.*

Our solar system is so vast it is almost impossible to express its distances in anything approaching human terms. We can only begin to understand its scale by using a variety of analogues. For instance, our moon is about 238,606 miles (384,000 kilometers) from Earth and it takes light traveling at 186,411 miles per second (300,000 km/sec) about 1.5 seconds to travel this distance. But to cross the solar system from the sun to the orbit of Neptune takes just over four hours at the speed of light. The New Horizons spacecraft, launched in 2006 with a speed of 36,661 mph (59,000 km/h) traveled from Earth to the lunar distance in just under seven hours, but took nine years to reach Pluto. Meanwhile, the radio signals from the spacecraft took 4.5 hours to reach Earth. This stupendous scale, and the vastness of interplanetary space, provides us with insurance that we will not be frequently struck by comets and asteroids that could end life on Earth. It also protects us from the gravitational influences of Jupiter, which could easily alter Earth's orbit and cause enormous climate changes hostile to life as we know it. But it does make human space travel difficult.

Our solar system consists of eight major planets: Mercury, Venus, Earth, Mars, Jupiter, Saturn, Uranus, and Neptune; together with five dwarf planets: Pluto, Ceres, Eris, Makemake, and Haumea. Many of these bodies also have satellites that orbit them; some as large as planets in their own right. Although Mercury and Venus have no satellites, Earth has its moon; Mars has Deimos and Phobos; Jupiter has no fewer than 75 satellites; Saturn has 82 satellites; Uranus has 27 and Neptune has 14. Even the dwarf planets have their own retinues, with Pluto having five; Haumea has two; and Eris and Makemake have one each.

Together with the major planets and dwarf planets, there are literally millions of asteroids and comets that orbit the sun in often intersecting paths. Two major belts exist between the orbit of Mars and Jupiter, called the Asteroid Belt, and between Neptune and Pluto called the Kuiper Belt. The Asteroid Belt contains over 100,000 known objects between a few feet and over 62 miles (100 kilometers) in size, generally made of rocky materials. The more-distant Kuiper Belt contains over 300 known objects larger than 31 miles (50 kilometers) and made primarily of ices. These objects remain in more-or-less stable orbits centered on the sun, but

can be perturbed by Jupiter and distant Neptune. Asteroid Belt objects can be diverted into orbits that cross the inner planets, producing potential impact hazards. Meanwhile, the Kuiper belt objects can be perturbed into orbits that cross the inner solar system, producing what we observe as periodic comets. When the solar system was young, there were far more of these orbit-crossing asteroids. Many of them impacted the surfaces of the inner plants as well as their moons, leaving behind a record of craters spanning a wide range of sizes. Many others were ejected by Jupiter and the outer plants into more distant orbits in the so-called Oort Cloud. In some sense, the planet-building process is still occurring, with Earth still continuing to gain about 56,000 tons (50,800 tonnes) each year.

Above *The scale of the outer solar system is vastly different than the crowded inner solar system where the terrestrial planets orbit. At the limits of our solar system beyond the orbit of Neptune we encounter a second "asteroid belt" called the Kuiper Belt, consisting of thousands of icy objects from 62 to 620 miles (100 to 1,000 kilometers) in size.*

						COMPOSTION		
Name	**Distance (AU)**	**Period**	**Rotation**	**Size (Earth=1)**	**Moons**	**Core**	**Mantle**	**Atmosphere***
Mercury	0.38	87.9d	58.6d	0.38	0	Fe	Silicates	Sodium
Venus	0.72	224.7d	243.0d	0.95	0	Fe/Ni	Silicates	Carbon Dioxide
Earth	1.00	365.3d	23.9 h	1.00	1	Fe/Ni	Silicates	Nitrogen, Oxygen
Mars	1.52	686.9d	24.6h	0.53	2	Fe/Ni	Silicates	Carbon Dioxide
Jupiter	5.20	11.8y	9.6h	11.21	75	Silicates	Hydrogen	Hydrogen
Saturn	9.54	29.5y	10.5h	9.45	82	Silicates	Hydrogen	Hydrogen
Uranus	19.19	84.1y	17.2h	4.00	27	Silicates	Ices	Hydrogen
Neptune	30.07	164.8y	16.1h	3.88	14	Silicates	Ices	Hydrogen

*Dominant

// Mercury

This innermost planet orbits the sun at an average distance of about 22 million miles (35 million kilometers) and travels once-around in 88 days. From this location, the disk of our sun would be three times larger than what we see from Earth. There have been many searches for additional planets in this inner zone, as well as any moons for Mercury, but none have ever been discovered. Because its orbit is very elliptical, and it feels the gravitational effects of the other planets, especially Jupiter, over the course of the next 5 billion years there is a chance that it may actually collide with the planet Venus. Because it is so close to our sun, the gravitational field of the sun actually warps the

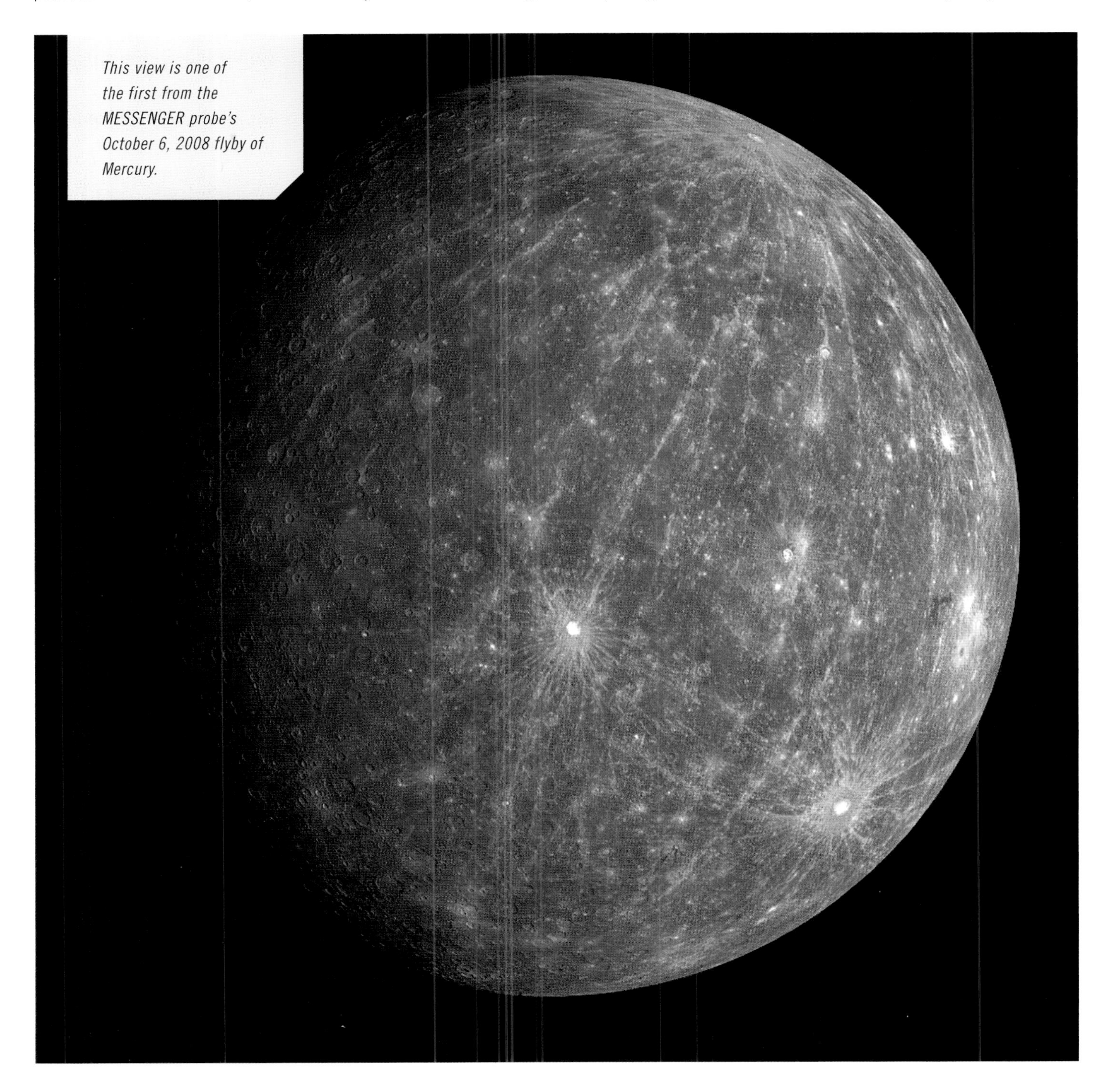

This view is one of the first from the MESSENGER probe's October 6, 2008 flyby of Mercury.

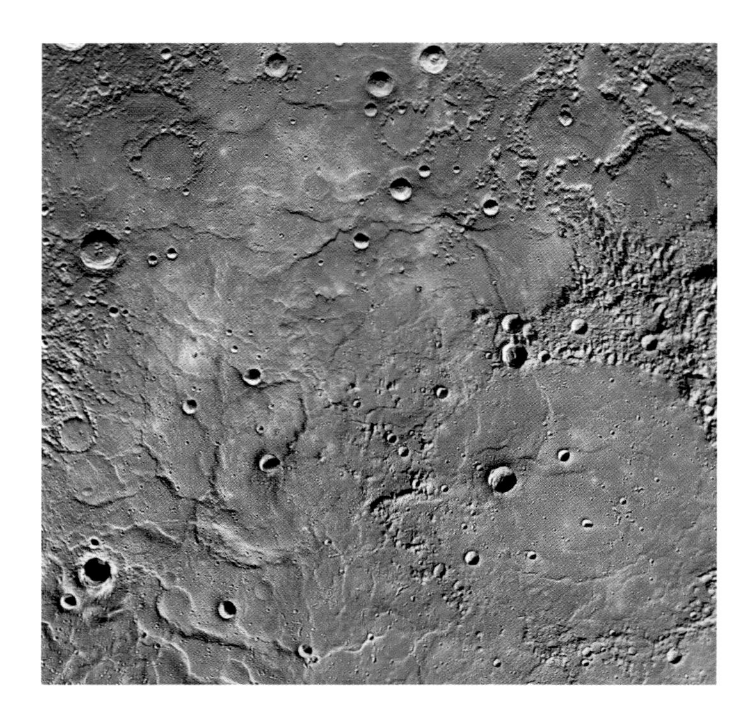

Above *A view of Mercury's northern volcanic plains is shown in enhanced color to emphasize different types of rocks on Mercury's surface. In the bottom right portion of the image, the 181-mile- (291-kilometer-) diameter Mendelssohn impact basin appears to have been once nearly filled with lava.*

geometry of the space in which Mercury travels. This causes a measurable effect called the precession of the perihelion that was first explained by Albert Einstein's general theory of relativity in 1915 and is considered one of the theory's most significant confirmed predictions.

Mercury has an iron core taking up nearly half the interior volume of this moon-sized planet. Its liquid-iron currents in the outer core zone would produce the strongest magnetic field of any planet in the inner solar system were it not for the planet's sluggish rotation period of 56 days. Without brisk electrical currents, its magnetic field is a thousand times weaker than Earth's, making it ineffective in diverting the harsh solar winds. Over billions of years, any atmosphere it started with due to volcanic outgassing would have been stripped away by the solar wind without this magnetic shield, leaving behind the wrecked, cratered world seen by spacecraft. The steady cooling of the iron core has also caused the entire planet to contract. The crackled crust left behind on the surface from this 3 to 6 mile (5 to 10 kilometer) shrinkage can be seen everywhere and this process appears to be continuing today.

Unlike our Earth with its 26.5° axis tilt, Mercury's rotation axis tilt is less than 1°. At the poles, the sun never gets more than about a half-degree above the horizon, so the poles are in near perpetual darkness. Because the angular speed of rotation of Mercury is similar to its orbital angular speed when the planet is closest to the sun, this leads to an interesting double sunrise and sunset during a single eight-day period. An observer would be able to see the sun peek up a little more than two-thirds of the way over the horizon, then reverse and set before rising again.

The planet's entire surface is heavily cratered in a way similar to other solid bodies in the inner solar system. Most of these impacts happened during what geologists on Earth call the Heavy Late Bombardment era, which ended 3.8 billion years ago. Just as Earth was struck by a large planetoid that produced our moon, Mercury suffered a massive impact that formed the 923-mile (1,500-kilometer) Caloris basin. The shock waves of the 124-mile (200-kilometer) impactor traveled through the interior of the planet and caused the eruption of chaotic terrain on the opposite, antipodal side of Mercury.

The surface temperature of Mercury swings from 806°F (430°C) at local noon, to 356°F (180°C) at local midnight, but even so, this is not the hottest planet in the solar system. For that we have to visit Venus. Remarkably, despite its high temperatures, there are some craters in the north polar regions that are in permanent shadow, and in which water ice has been detected. The amount is in no way comparable to Earth's vast icecaps but nevertheless amounts to perhaps 2.4 cubic miles (10 cubic kilometers).

Below *A region about 100 miles (160 kilometers) across, near Mercury's equator.*

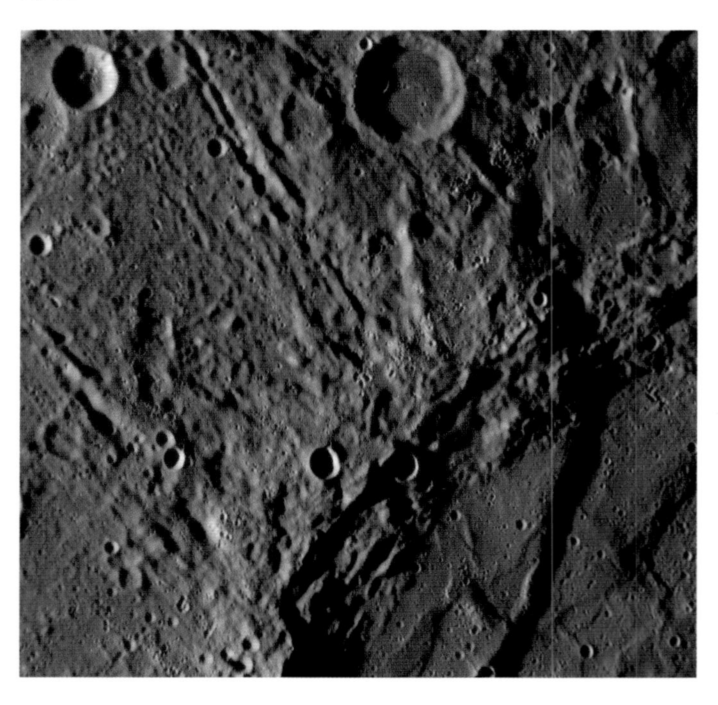

// Venus

For centuries very little was known about this planet, which orbits the sun every 235 days from a distance of 67 million miles (108 million kilometers). Even though it makes its closest approach to Earth at a distance of about 25 million miles (40 million kilometers), through the most powerful telescopes, it appears as a completely blank, featureless disk. This disk, however, goes through various phases of illumination from full to crescent because it is always located inside the orbit of Earth. Occasional glimpses of clouds in the atmosphere allow astronomers to calculate a rough rotation period of 224 days, but this rotation is in the opposite direction (clockwise) from all other planets. Venus has by some means flipped upside down over the course of its history so that, from its surface, the sun rises in the west and sets in the east.

Venus is nearly the same size and mass as Earth. It even has a dense atmosphere, which may well have been very Earth-like billions of years ago. There may also have been large planetary oceans not unlike our own. But its active volcanism has released far too much carbon dioxide into the atmosphere. Its location at the inner edge of our sun's habitable zone, together with a dense carbon dioxide atmosphere, caused its runaway greenhouse heating to eventually evaporate its oceans, leaving behind a rocky surface heated to a relentless temperature of 860°F (460°C). Science fiction authors as recently as the 1950s expected Venus to be a swampy, hot, and humid world suitable for human colonization. In 1956, astronomers using radio telescopes measured the surface temperature of Venus for the first time and discovered it to be incredibly hot: over 572°F (300°C). Spacecraft such as Mariner 2 confirmed this temperature in 1962, and that pretty much spelled the end of an entire science fiction genre. Only lunatics—or scientists—would ever dare to visit the surface of this world, whose actual name should more suitably be Crematoria or Hades.

The atmosphere contains far more than just 96 per cent carbon dioxide and 3 percent nitrogen. Sulfur released through volcanism combined with the water vapor in the atmosphere to create dense clouds that are rich in sulfuric

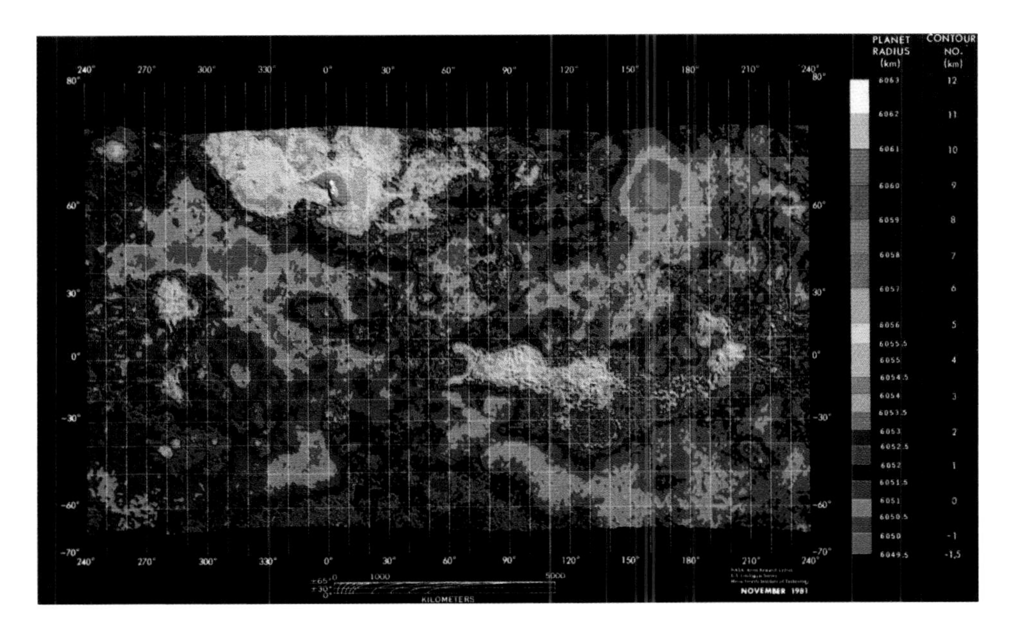

Above *NASA's Mariner 10 spacecraft captured this seemingly peaceful view of a planet the size of Earth, wrapped in a dense, global, and utterly lethal cloud layer.*

Left *A topographic map of Venus based on data obtained by NASA's Pioneer Venus Orbiter.*

Right *This image is a composite of radar surface data from NASA's Magellan spacecraft and the Pioneer Venus Orbiter.*

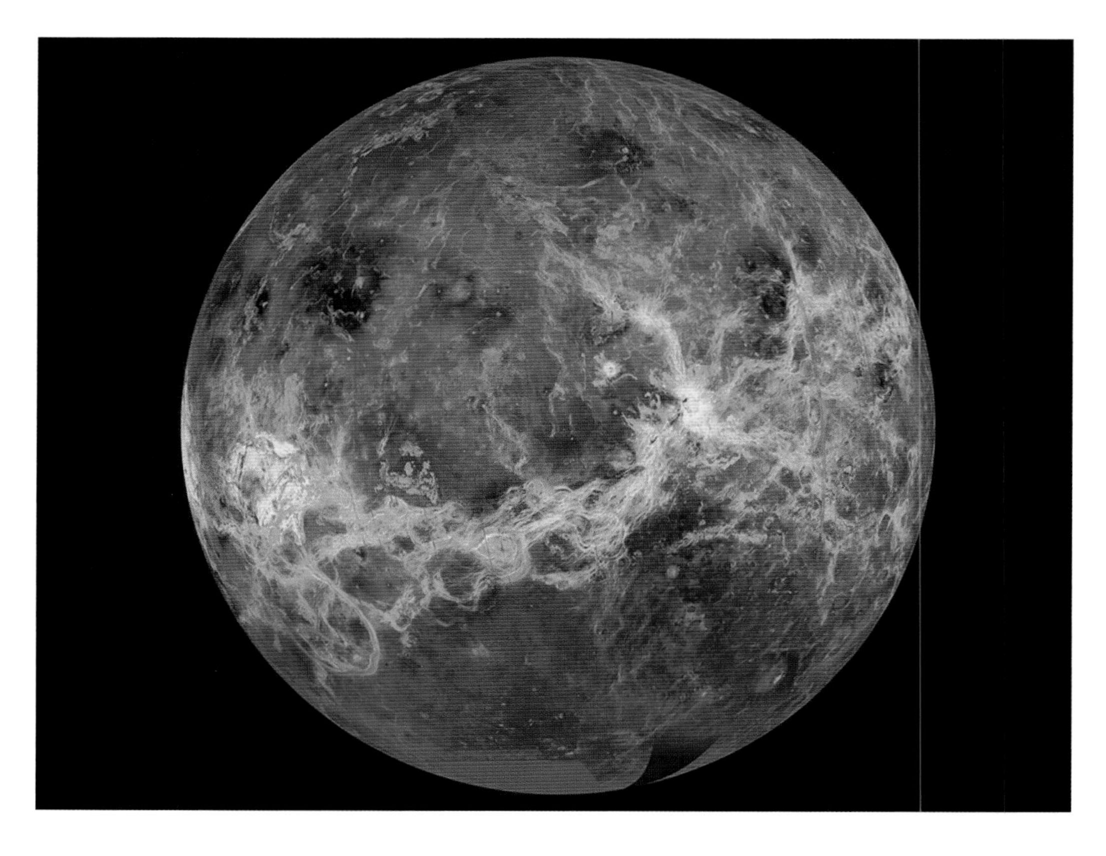

Below *A colorized photo of the surface of Venus taken by the Soviet Venera 13 spacecraft on March 1, 1982.*

acid. The rain from these clouds constantly falls through the lower atmosphere as a highly corrosive agent for any spacecraft that attempts to reach the surface. At a pressure over 90 times that of Earth, if a spacecraft is not corroded or baked into uselessness by the time it reaches the surface, it is surely crushed by the sheer weight of the atmosphere itself. Of all the biblical descriptions of Hell, this is surely its most apt location.

Between 1975–82, the Soviet Venera spacecraft repeatedly landed on the surface of Venus and returned brief glimpses of a close-in landscape marked by dark, slate-like rocks that appeared weathered and smooth. Due to the extreme heat and using conventional technology, these probes even with their active refrigeration systems lasted less than a few hours before failing. Basically, it would be like putting your cell phone inside your kitchen oven and expecting it to keep operating for several weeks. Modern designs for future landers will use electronic circuitry that can survive temperatures as high as 932°F (500°C) based on silicon carbide rather than pure silicon.

// Earth

Entire libraries are filled with books that describe this blue ball world that is unique in all of the solar system. Its "just-right" combinations of temperature, magnetic field, and water were able to trigger the emergence of life, which then evolved to become beings capable of writing this book. Many things had to come together to allow sentient humans to appear, including a fateful asteroid 65 million years ago that eliminated our dinosaur competitors.

As an astronomical object, it has a radius of 3,963 miles (6,378 kilometers) and is virtually identical to Venus in its internal structure although, with a rapid rotation period of 24 hours, its vigorous core currents generate the most powerful magnetic field of all the rocky bodies in the solar system. This field diverts the solar wind and fast-moving clouds of solar plasma so that they do not directly collide with our atmosphere. This magnetic shield has protected our atmosphere from eroding away and turning our planet into another Mercury. Its thin, silicate-rich crust is fragmented into continental plates that move, collide, and reform due to convection currents in the mantle. If it had a thicker crust and no lubricating ocean water there would be little or no continental drift and volcanism produced by the subducting plates. Over the last 3 billion years there have been several supercontinents that have formed and broken up. The most recent one is Pangaea, which broke up 250 million years ago and led to the current continental configuration.

Comfortably located in the habitable zone, Earth's evolution diverged from that of ill-fated Venus because volcanism was not as lavish and liquid water succeeded in burying billions of tons of carbon dioxide every year, preventing a greenhouse catastrophe. The biggest difference between Earth and its neighboring planets is its nitrogen- and oxygen-rich atmosphere. This is actually the planet's third atmosphere. The primordial atmosphere of methane, ammonia, and carbon dioxide was replaced by volcanism and outgassing and remained methane- and carbon dioxide-rich until about two billion years ago, when oceanic organisms began respirating oxygen and steadily increased the oxygen supply to 22 percent. This set the stage for the appearance of complex oxygen-loving organisms.

Due to continental drift, the changing orbit of Earth, and its atmospheric "green house" constituents from volcanic outgassing, Earth has gone through many periods of ice ages and glaciations. Around 650 million years ago, Earth was nearly covered by glaciers and ice caps called the Snowball Earth era. This was replaced by 100-million-year ice age periods where the polar caps expanded and retreated. Ice ages also consist of smaller-duration glaciation episodes lasting a few hundred thousand years. The last glaciation episode, called the Quaternary glaciation, ended about 15,000 years ago at the start of the Eemian Interglacial Period, and the milder climates set the stage for a huge increase in the human population and technological inventiveness.

Right *This spectacular "blue marble" image is based on the most detailed collection of true-color imagery of the entire Earth to date. Using a collection of satellite-based observations, scientists and visualizers stitched together months of observations of the land surface, oceans, sea ice, and clouds into a seamless, true-color mosaic of every square mile of our planet.*

Below *Earth compared to the sizes of the other planets in our solar system.*

// Mars

This dusky-red planet has been the stuff of mythology for millennia, usually evoking anger and warfare even in the 20th century when hostile Martians invaded Earth. Modern science fiction writers, inspired by earlier, and incorrect, observations by the Italian astronomer Giovanni Schiaparelli of water-filled canals on Mars, have envisioned ancient and wise Martian civilizations, or extensive colonization by humans. Recent stories have turned toward the first attempts in this century of humans traveling to Mars and setting up the first fledgling scientific outposts. Even entrepreneurs such as Elon Musk at SpaceX have pledged their talents and resources to the first manned trips to Mars by the end of the 2020s. We now know enough about the effects such three-year expeditions would have on humans and their physiology to know that the risk of some deaths and colony collapse is not inconsequential.

Sadly, this is a planet hovering on the outskirts of the habitable zone for which its size circumscribed its destiny. With too feeble a gravity to hang onto its atmosphere against the heating by the sun, and with too feeble an interior current to maintain a strong magnetic field, its atmosphere leaked away many billions of years ago. Left behind were the fossil geologies of rivers and oceans that once existed, but whose basins are now occupied by blowing dust and craters. Being the nearest planet where humans can land and survive, dozens of spacecraft have traveled to it and many have landed, returning tantalizing views of a barren but surprisingly Earth-like surface. Subsurface deposits of permafrost rich in water ice beckon future explorers, human and robotic, to search for traces of microbial fossils or even life existing today.

Mars orbits the sun every 687 days at an average distance of 141 million miles (227 million kilometers), and has a near-Earth-like day length of 24.6 hours. Its surface gravity is about one-third of Earth so colonists would have a definite bounce to their step. The atmosphere is so thin at an equivalent Earth altitude of about 28 miles (45 kilometers) you would need a full spacesuit, and not a simple face mask and warm clothing as many older science fiction stories suggested. Every 780 days it is closest to Earth at distances of about 22 million miles (35 million kilometers). These opposition windows are the target dates for both unmanned and manned expeditions to Mars. So far, spacecraft have been sent to Mars each opposition since the early 1960s. Typical transit times are about 224 days with chemical rockets, which presents a severe challenge for human spaceflight.

Since the 1960s and by 2020, 62 missions have traveled, or attempted to travel, to Mars. Of these, 14 have taken up orbits, and ten have successfully landed rovers, like Opportunity, Spirit, Sojourner and fixed laboratories such as Viking and Insight, that returned views of the surface or other scientific data. The most extensive of these missions has been NASA's Curiosity rover,

Below *Mars as seen from orbit by NASA's Viking mission.*

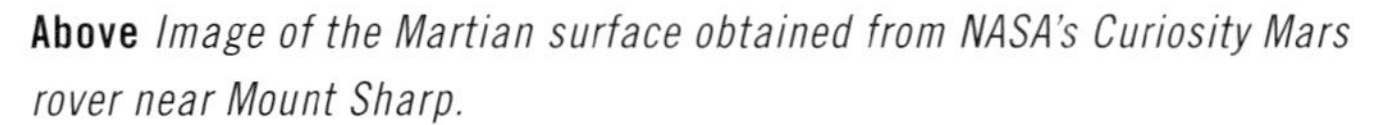

Above *Image of the Martian surface obtained from NASA's Curiosity Mars rover near Mount Sharp.*

which arrived in 2012 and currently has traversed nearly 14 miles (23 kilometers) in the vicinity of the Gale crater. The Curiosity rover and others of its kind have confirmed many scientific conjectures about the evolution of the Martian surface and especially its history of liquid water. On 18 February 2021 the Perseverance rover and its helicopter called Ingenuity arrived on the Martian surface.

It will continue Curiosity's research and may hopefully find actual traces of organic molecules and perhaps even fossils. Over 300 chemical compounds are known to be unique features of organic life. The discovery of even a handful of these would be a dramatic moment in human history. The Perseverance rover may also discover microfossils embedded in the ancient clays of the Jezero crater.

Spacecraft images from orbit have for decades revealed landforms that indicate the past existence of substantial quantities of running water. River deltas and a vast Northern Hemisphere ocean basin are clearly visible from orbit. The discoveries made by landers and rovers essentially confirmed the existence of water, not only as an ancient liquid element of the Martian surface but as a current icy feature of its subsurface geology. Unfortunately, humans missed the heyday of a verdant Mars with flowing water and a thick atmosphere by about 3 billion years. Given how quickly life appeared on Earth within its first billion years, there is enormous cause for optimism for finding biological microfossils in some rocky stratum fed by long-vanished rivers.

Below *The Perseverance rover and the Ingenuity helicopter are the latest visitors to Mars following their 18 February 2021 arrival. This 'selfie' image was taken by Perseverance and shows Ingenuity and a portion of the local landscape.*

// Jupiter

Jupiter orbits the sun at an average distance of 483 million miles (778 million kilometers) and takes 11.8 years to complete one "year." Its 10-hour rotation period with a diameter 11 times that of Earth means that it has the fastest rotation speed of any other planet, with its outer equatorial layers traveling at 28,000 mph (45,000 km/h). If you took all of the planets and combined their mass you would equal just over one-third the mass of Jupiter. With its retinue of 75 satellites, four of which are as large as the planet Mercury, it resembles some odd kind of diminutive solar system rather than a member of our own. Had it been 50 times more massive, Jupiter would have been able to fuse deuterium in its core to become a brown dwarf star. Even today, the gradual contraction of its core at a rate of ¾ inch (2 centimeters) per year produces enough heat that Jupiter radiates more energy than it receives from the sun.

The intense radiation belts that surround Jupiter are a signpost of one of the strongest magnetic fields in our solar system, excepting that of the sun itself. This field originates deep inside the planet in a dense and fast-moving current of hydrogen atoms compressed into a metallic state that overlays what may be a rocky core. This metallic hydrogen zone extends about 78 percent of the planet's radius to its surface. At the tremendous internal pressures and temperatures of only about -400°F (-240°C), another zone composed of liquid hydrogen extends to within 620 miles (1,000 kilometers) of its visible surface followed by an "atmosphere" of gaseous hydrogen

Below *Jupiter shines like a jewel in the night, a giant pillar of color and power that towers above everything in the solar system besides the sun itself.*

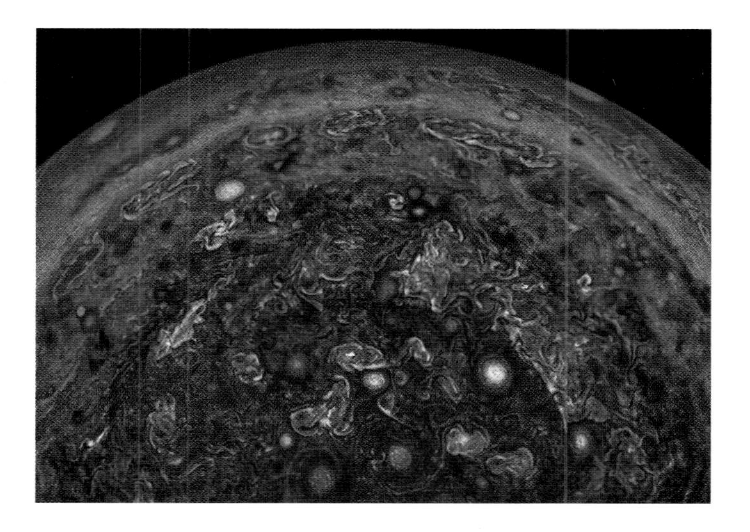

Above *NASA's Juno spacecraft soared directly over Jupiter's south pole and acquired this image from an altitude of 62,000 miles (100,000 kilometers) above the cloud tops.*

and helium. Its atmosphere whirls around at speeds over 497 mph (800 km/h) with clouds stretched into numerous belts and cyclonic storms driven by the convecting and shearing atmosphere. Like our sun, it is composed of 75 percent hydrogen and 24 percent helium by mass. The dramatic coloration of its visible clouds and other surface features is the result of phosphorus, sulfur, and a variety of organic compounds called tholins. Its Great Red Spot, first observed telescopically around 1665, is the largest giant cyclone among many other similar features that arise due to the complex atmosphere dynamics. It has been steadily decreasing in size from about seven times the diameter of Earth in the 1800s, to its present size of just over twice the diameter of Earth. At this rate, it may vanish entirely by *c.*2035.

There have currently been seven flyby missions of this planet beginning with Pioneer 10 in 1973 and more recently the New Horizons mission in 2007. The Voyager 1 spacecraft made the first discovery of Jupiter's ring system in 1979. These dust rings exist within the orbits of its inner moons Adrastea, Metis, Amalthea and Thebe and are located between 56,000 and 140,000 miles (90,000 and 226,000 kilometers) from the planet's center and are probably produced by ejected material from the moons themselves. The rings are constantly losing mass and would disappear in under 1,000 years, so they must be constantly replenished, but the details for how this happens remain unknown.

The Galileo mission, which arrived in 1995, and the Juno mission arriving in 2016, orbited the planet and provided a wide variety of investigations of its atmosphere, radiation

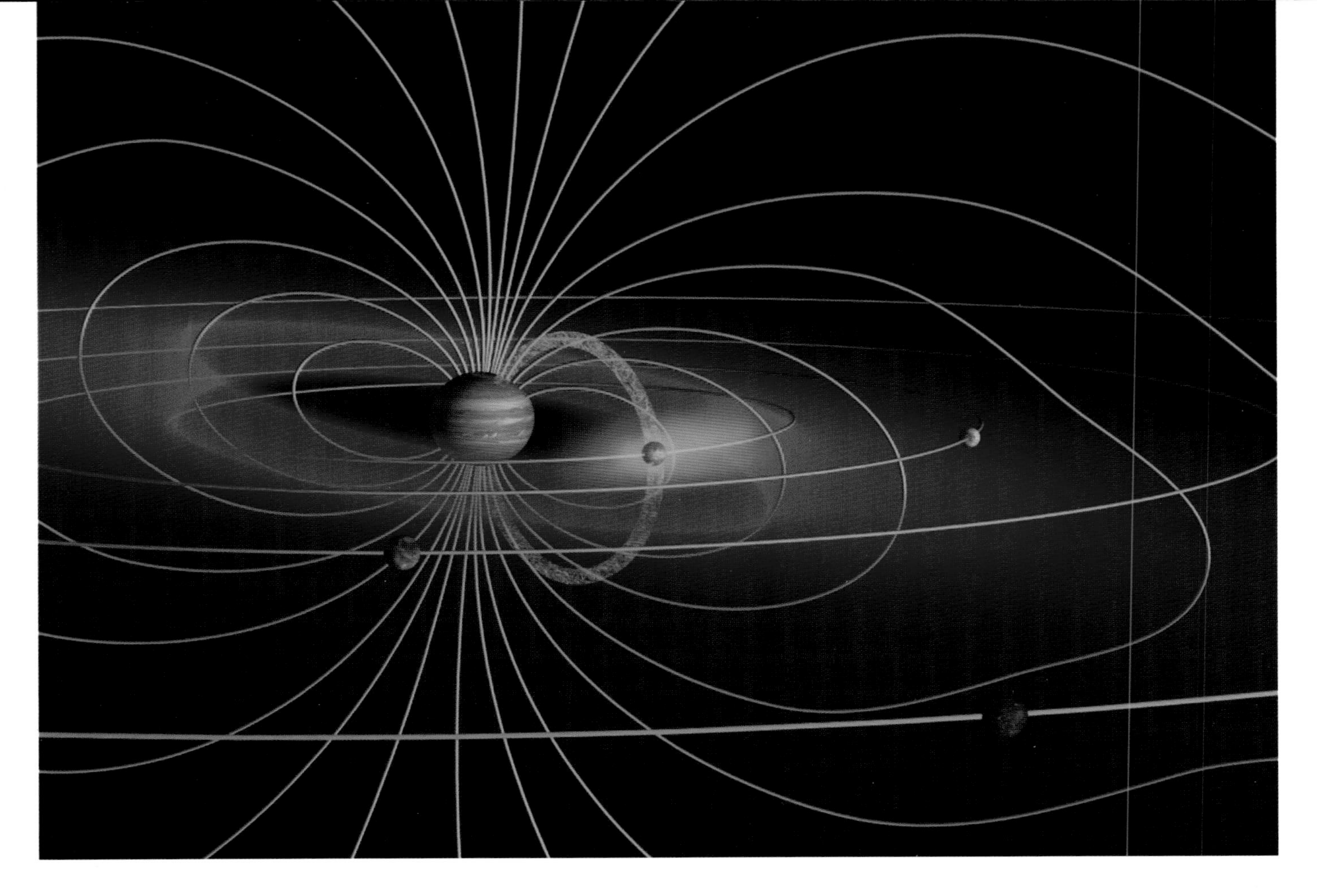

Above *Schematic drawing of Jupiter's magnetic field encompassing its four moons. The intense radiation is lethal to humans and most spacecraft without enormous amounts of shielding.*

belts and imagery of its moons. In 2022 and 2023, ESA will launch the Jupiter Icy Moons Explorer and NASA will launch the Europa Clipper, which will explore in detail the surfaces of Ganymede, Europa, and Callisto and investigate what are inferred from surface features to be their sub-surface liquid oceans. Among all the bodies in the solar system, excepting Earth, Europa, Ganymede, and Callisto remain the most likely places to find life, buried deep within their sub-surface oceans. These oceans are overlain by a thick 6.2 to 62 miles (10 to 100 kilometers) crust of solid ice, so reaching these oceans poses a huge challenge for future robotic missions seeking direct contact with life signs.

Below *A view of Jupiter's tumultuous clouds.*

// Saturn

Left *A view of Saturn in 2019 from NASA's Hubble Space Telescope captures exquisite details of the ring system.*

This is the second of two gas giants in the solar system, and orbits the sun every 29.5 years at a distance of just under 932 million miles (1.5 billion kilometers). Its 10.3-hour rotation period is only slightly longer than Jupiter's. Its mass is 95 times that of Earth or just under one-third the mass of Jupiter. Its diameter is about nine times that of Earth. Like Jupiter, it has a mixed rocky and metallic hydrogen core, surmounted by a liquid hydrogen mantle and a gaseous outer atmosphere. The currents in the metallic hydrogen core produce a magnetic field about 5 percent the size of Jupiter's. Insufficient ultraviolet light from the sun reaches its cloud tops to produce the abundant, colorful compounds found on Jupiter that reveal its dramatic atmosphere features. Saturn's atmosphere and clouds are more difficult to detect but nevertheless show similar banding and cyclonic storms at a lower contrast. Its slightly smaller size but similar rotation rate, as for Jupiter, produces atmospheric speeds of 1,118 mph (1,800 km/h).

Saturn more than makes up for its smaller size compared to Jupiter by its dramatic rings, which extend from about 4,350 to 74,565 miles (7,000 to 120,000 kilometers) from the planet's cloud tops. With a thickness of only 66 feet (20 meters), this ring system is the largest and thinnest dynamically stable object in our solar system. Once thought to be solid at the dawn of telescopic observations, even flyby spacecraft such as Cassini have been unable to resolve the texture of these rings as individual dust grains or icy fist-sized rocks. Some of the moons such as Pandora and Prometheus act to shepherd the ring particles into confined orbits that prevent the

Below *A comparison of the interiors of the gas giant planets Jupiter and Saturn revealing their differing compositions.*

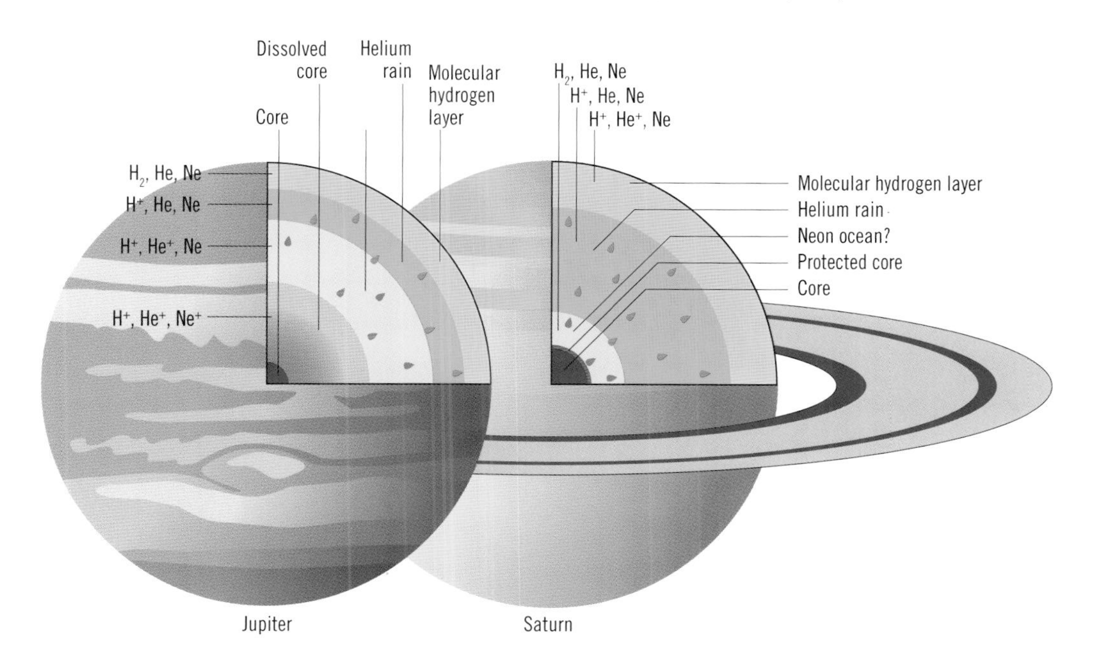

dispersal of the rings. Whether the rings were material left over from the formation of Jupiter, or are material from one or more destroyed moons, cannot be determined at this time.Earth was bombarded by more planetesimals, slowly enlarging further and further.

The Pioneer 11 and Voyager 1 and 2 spacecraft flew by Saturn between 1979 and 1981, followed in 2004 by the Cassini/Huygens spacecraft that went into orbit. The Voyager spacecraft produced the first high-resolution images of Saturn and some of its moons, which greatly exceeded the ability of ground-based telescopes to discern. The high-resolution images of the ring system from many orientations provided crucial data for understanding their composition. In 2006, Cassini discovered water geysers on Saturn's moon Enceladus and continued the Voyager imaging and mapping of many of its moons. In particular, imagery of the rings revealed complex gravitational interactions between moonlets and ring material such as waves and direct evidence of shepherding activity. There are currently no funded programs to return to Saturn.

Above *The Cassini spacecraft captured this image of a small object (center) in the outer portion of Saturn's B ring, casting a shadow on the rings. This new moonlet, seen near the center of the image, was found by detection of its shadow which stretches 25 miles (41 kilometers) across the rings, indicating a moonlet size of about 1,312 feet (400 meters) across.*

Below *A composite image of auroras above Saturn's south pole region using the Hubble Space Telescope.*

// Uranus

Above *An image of the planet Uranus, taken by the spacecraft Voyager 2 in 1986.*

The first of the ice giants, this planet is 5 percent the mass of Jupiter and about four times the diameter of Earth. Located 1.8 billion miles (2.9 billion kilometers) from the sun, it takes 84 years to orbit the sun. Its outer atmosphere temperature below -370°F (-224°C) receives only 0.25 percent the warmth from the sun as Earth does. Although it rotates once every 17.3 hours, the rotation axis is very close to being along the plane of its orbit, so it almost literally rolls around the solar system. This makes for the unique seasonal changes found nowhere else in the solar system. Each pole gets about 42 years of continuous sunlight followed by 42 years of darkness as each pole is alternately pointed toward the sun and away, similar to Earth's solstices. During the equinoxes, the sun shines directly over the equator and each hemisphere gets an almost

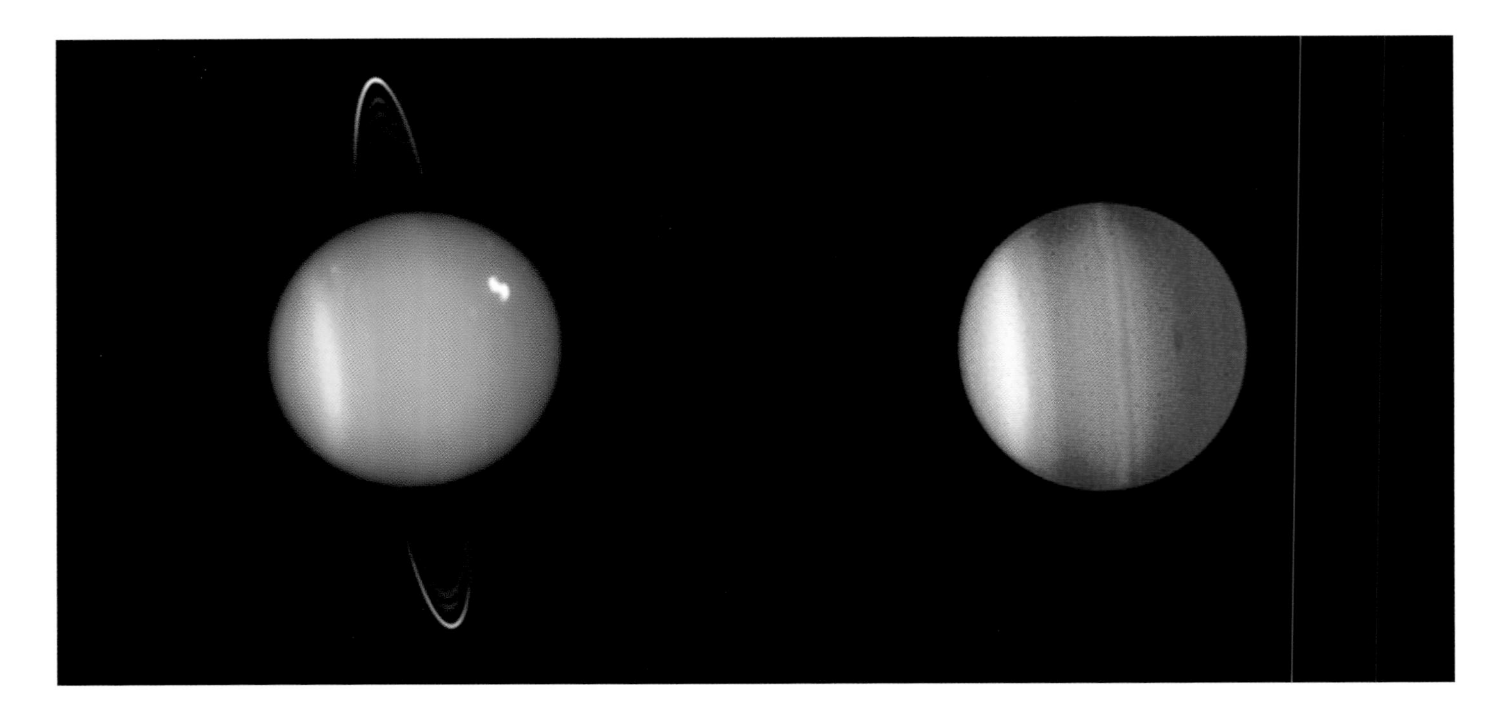

Earth-like day-night cycle. The reason for this extreme axial tilt may have been a huge impact close to the time of the planet's formation since the orbits of its moons are also in the planet's equatorial plane and show no signs of a recent disturbance or major planetary impact.

Uranus has a complex ring system consisting of 13 individual components, probably composed of icy dust particles, which are a common ring component among the outer planets. The rings appear to be quite young and probably did not form with the planet like Saturn's rings. A possibility is that they are particles left over from moonlet collisions or material ejected from the surfaces of its existing moons such as Mab, which orbits just outside the outermost ring.

The interior of Uranus has a deep icy layer occupying two-thirds of the planet, surmounted by a thick hydrogen/helium atmosphere with traces of methane gases. Its deep core could be an Earth-sized, silicate and iron-rich rocky object. At the temperatures of the interior, some portion of the icy interior may even be a partially liquid ocean of methane and water. Some models even suggest that methane molecules would be fractionated into pure carbon, which would rain down as diamonds on a liquid diamond ocean at the base of the mantle. Its featureless appearance and color belie the fact that methane absorbs some of the visible light reflecting from the cloud tops and renders a color similar to aquamarine or cyan. Faint details can be seen at other infrared wavelengths revealing belts of fast-moving clouds resembling those of Jupiter or Saturn but with less detail.

Right *Combination of images from the Hubble Space Telescope and the Voyager 2 probe. In 2012 and 2014 powerful bursts of solar winds caused Uranus' aurora to flare.*

Above *These Hubble Space Telescope images show the varied faces of Uranus.*

Uranus has been visited by the Voyager 2 spacecraft in 1986 and produced high-resolution images of its five large moons along with the discovery of an additional ten smaller ones not visible from Earth. It also studied nine of its rings and discovered two additional ones. A number of missions have been proposed to follow-up on the Voyager studies but none are currently scheduled by NASA or ESA for the launch windows of 2030 and 2034.

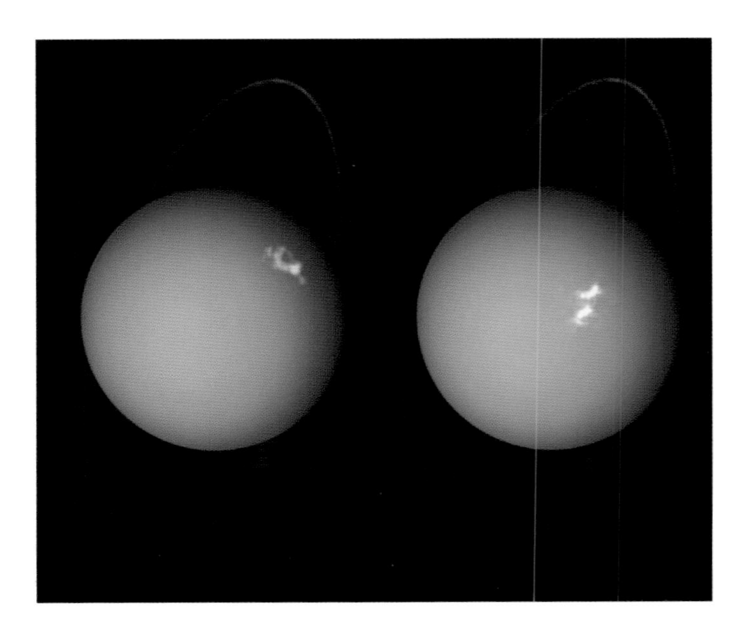

// Neptune

The most distant planet in our solar system orbits 2.8 billion miles (4.5 billion kilometers) from the sun and takes just under 165 years to complete a trip around the sun. It is 3.9 times the size of Earth, 5.5 percent the mass of Jupiter and rotates once every 16.1 hours. A near-twin to Uranus in size and composition, its Earth-sized rocky core is surrounded by a thick ocean of liquid water, ammonia, and methane. The pressure and temperature are believed to be high enough near the core-mantle boundary that, as for Uranus, the carbon-rich methane may even crystallize into solid diamond rain that covers the rocky core. Its upper atmosphere, like Neptune, is hydrogen-rich with significant amounts of methane and ammonia, but its distinct blue color is the result of methane

Below *NASA's Voyager 2 image, taken at a range of 4.4 million miles (7 million kilometers) from the planet, shows the Great Dark Spot and its companion bright smudge.*

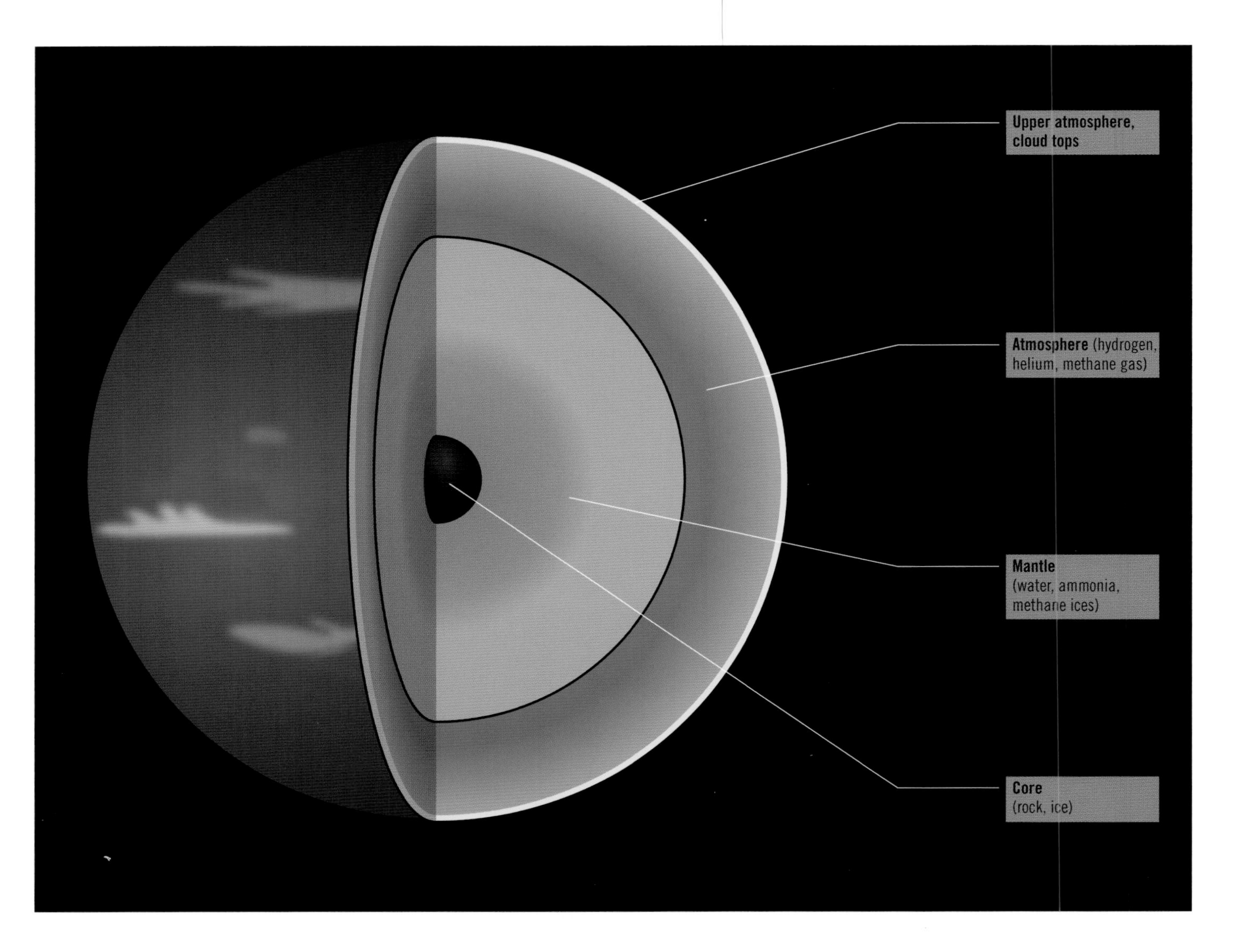

Above *Composition and interior structure of Neptune.*

molecules absorbing the red light in the solar spectrum and reflecting the blue light back into space. Uranus allows some of the green light in the spectrum to also be reflected, giving it a more greenish-blue color while most of the green light is absorbed by Neptune's atmosphere, giving it a more blueish color. Curiously, at its distance, sunlight is 0.1 percent as effective in heating the planet, nevertheless its outer atmosphere has a temperature of over 896°F (480°C). The reason for this high temperature on such a remote planet is not understood as yet

Unlike Uranus, Neptune's atmosphere is more transparent so details such as storms and other meteorology can be more easily seen. It has a large Dark Spot with winds up to 1,553 mph (2,500 km/h) that are the fastest in the solar system. Neptune's magnetic field is nearly 30 times stronger than Earth's, but instead of being closely aligned with its rotation axis, it is tilted by over 45° from the rotation axis. It is also offset from the center of the planet rather than being centered on it, as for Earth. Astronomers do not at this time know why these magnetic tilts exist.

In 1989 the Voyager 2 spacecraft flew by Neptune, making it the only spacecraft to have visited this world so far. At this extreme distance it took about four hours for the signals to reach Earth. High-resolution images were obtained for the moons Triton and Nereid as well as the discovery of six new moons. The spacecraft also investigated the rings of Neptune, discovering two additional rings, possibly composed of small icy or silicate grains. Studies of these rings at the Keck Observatory in 2003 show significant changes to them, suggesting that they are not stable features and may eventually vanish. Several spacecraft missions have been proposed for launches in the late 2020s or early 2030s but none have been funded to continue development.

// Dwarf planets

These large objects resemble ordinary planets because they orbit the sun and are massive enough that gravity has deformed them into a round shape. But there is one reason why they are different and represent a new category of solar system objects. Planets have completed their formation process and have swept out all of the material near their orbital paths. Dwarf planets, by contrast, are still embedded in the Asteroid or Kuiper Belts from which they are continuing to form. Over time, perhaps measured in billions of years, they may finally absorb the remaining rubble close to their orbital paths and become fully-fledged planets. When this definition was adopted by astronomers in 2006, Pluto was demoted to a dwarf planet and there was considerable public outcry.

The first two dwarf planets to be studied in detail were Ceres and Pluto, thanks to the spacecraft Dawn launched in 2007 and New Horizons launched in 2006. They arrived at Ceres and Pluto in 2015 and returned hundreds of high-resolution images, revealing surfaces of surprising complexity. Although as expected Ceres was heavily cratered, traces of white surface material identified as sodium carbonate in one of these large craters called Occator, revealed a vast sub-surface reservoir that could contain liquid water. For Pluto, its surface was a pastiche of landforms, methane glaciers and evidence of new formations, perhaps due to the gravitational tidal forces

Below *Enhanced color global view of Pluto, taken when NASA's New Horizons spacecraft was 280,000 miles (450,000 kilometers) away.*

Below *This New Horizons image, over 249 miles (400 kilometers) across, reveals ice convection cells and other very young features, including ice mountains over 1.2 miles (2 kilometers) tall*

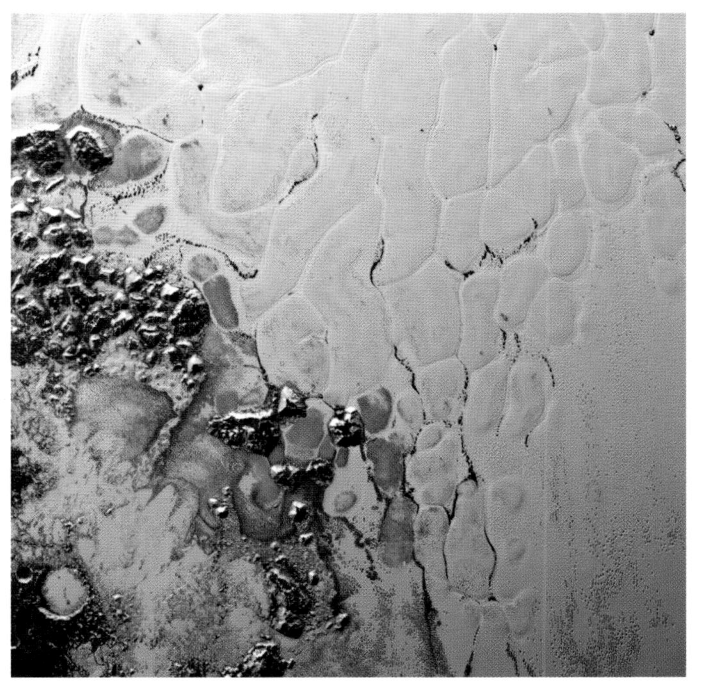

Name	Distance (AU)	Diameter (Moon=1.0)	Density (Water=1)	Rotation	Moons	Reflectivity
Pluto	39.5	0.68	1.8	6.1d	5	50%
Eris	67.8	0.67	2.5	25.9h	1	96%
Haumea	43.2	0.45	2.0	3.9h	2	66%
Makemake	45.6	0.41	1.7	22.8h	1	81%
Gonggong	67.4	0.35	1.7	22.4h	1	14%
Quaoar	43.7	0.31	2.0	17.7h	1	11%
Sedna	506.8	0.29	unknown	10.0h	0	32%
Ceres	2.8	0.27	2.2	9.1h	0	9%
Orcus	39.4	0.26	1.6	13.0h	1	23%

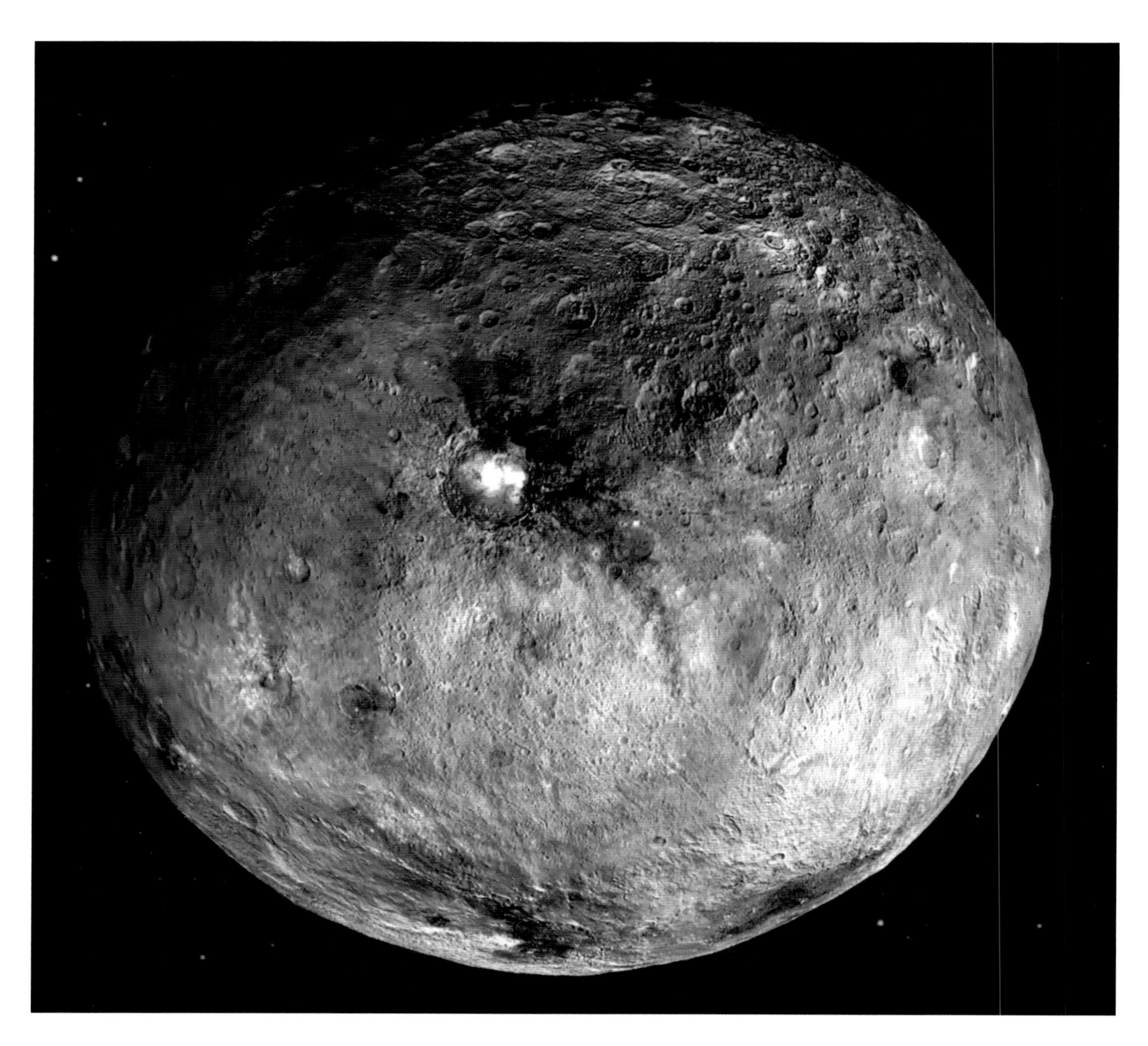

exerted by its large moon, Charon. It may also have an ocean of water many miles below its surface.

Little is known about the remaining dwarf planets other than their relatively small sizes and densities. They all appear to be slightly denser than water or water ices, so some amount of rocky material probably forms their cores. Their surfaces vary enormously in reflectivity, from nearly pure snow (Eris, Haumea, and Makemake) to worn asphalt (Ceres, Quaoar, and Gonggong). Some of the most distant objects such as Makemake have a distinct reddish tinge, which astronomers believe is due to the presence of molecules called tholins produced when material rich in methane is altered by exposure to solar ultraviolet light and cosmic rays.

Above *This false-color image shows the dwarf planet Ceres and the white sodium carbonate deposits at the center of Occator crater.*

The remaining dwarf planets have orbits even further away than Pluto. While Pluto once defined the edge of our planetary system some 4.7 billion miles (7.5 billion kilometers) from the sun, Makemake, Haumea and Eris orbit at 3.5, 4, and 6.3 billion miles (5.6, 6.4 and 10.1 billion kilometers). More remote dwarf planets may eventually be discovered.

// Planetary moons

The variety of planetary moons in our solar system is nothing less than bewildering and would require a book as large as the one you are reading to completely describe. They come in all sizes, shapes, and compositions from solid rock to solid ice. Many are distinguished by dramatic and even puzzling surface markings. Collectively, the eight planets in our solar system have 201 satellites, with more discovered every few years. Recent studies suggest that Jupiter alone may have as many as 600 additional moons smaller than a few hundred yards. Most of the larger moons were likely formed from the accreting material that built up the planets billions of years ago such as our own moon, or the large Galilean moons of Jupiter. But other moons may have been captured from the Asteroid Belt such as the satellites of Mars and the outermost moons of Jupiter and Saturn. Two new temporary moons to our own Earth were recently discovered, called 2006RH120 and 2020 CD3, which may orbit for a few years and then continue on their way. Were it not for the gravitational influences of our moon, these 3-feet-sized bodies may have become permanent moonlets.

The surface features of planetary moons are usually dominated by heavy cratering left over from the late bombardment era and numerous stray impacts over that last billion years. Some moons are large enough to have various degrees of surface renewal. Jupiter's moon Io has numerous sulfur volcanoes that resurface the moon every few million years. Other moons such as Charon (Pluto) and Europa (Jupiter) have crusts that deform and crack under the tidal action of their planet. Water geysers have been spotted on Neptune's moon Triton, and Saturn's moon Titan appears to have many active cryo-volcanoes that are spewing out liquid water and ammonia. Then there is the case of Miranda, a moon of Uranus with such a complex surface that some astronomers have proposed Miranda was smashed to pieces billions of years ago by a massive impact. The impact was not strong enough to completely eject the pieces. Gravity eventually pulled these pieces back together so that their differing geologies can now be seen as a dramatic hodge-podge of surface patterns.

Among the most intriguing worlds is Titan, the largest satellite of Saturn, which has an atmosphere about 50 percent denser than Earth's. Titan is 50 percent larger than Earth's moon, and has a surface gravity that is 11 percent of Earth's. Spacecraft such as Cassini took fleeting radar and infrared images of its surface during many orbits and

Below *The major moons of the solar system.*

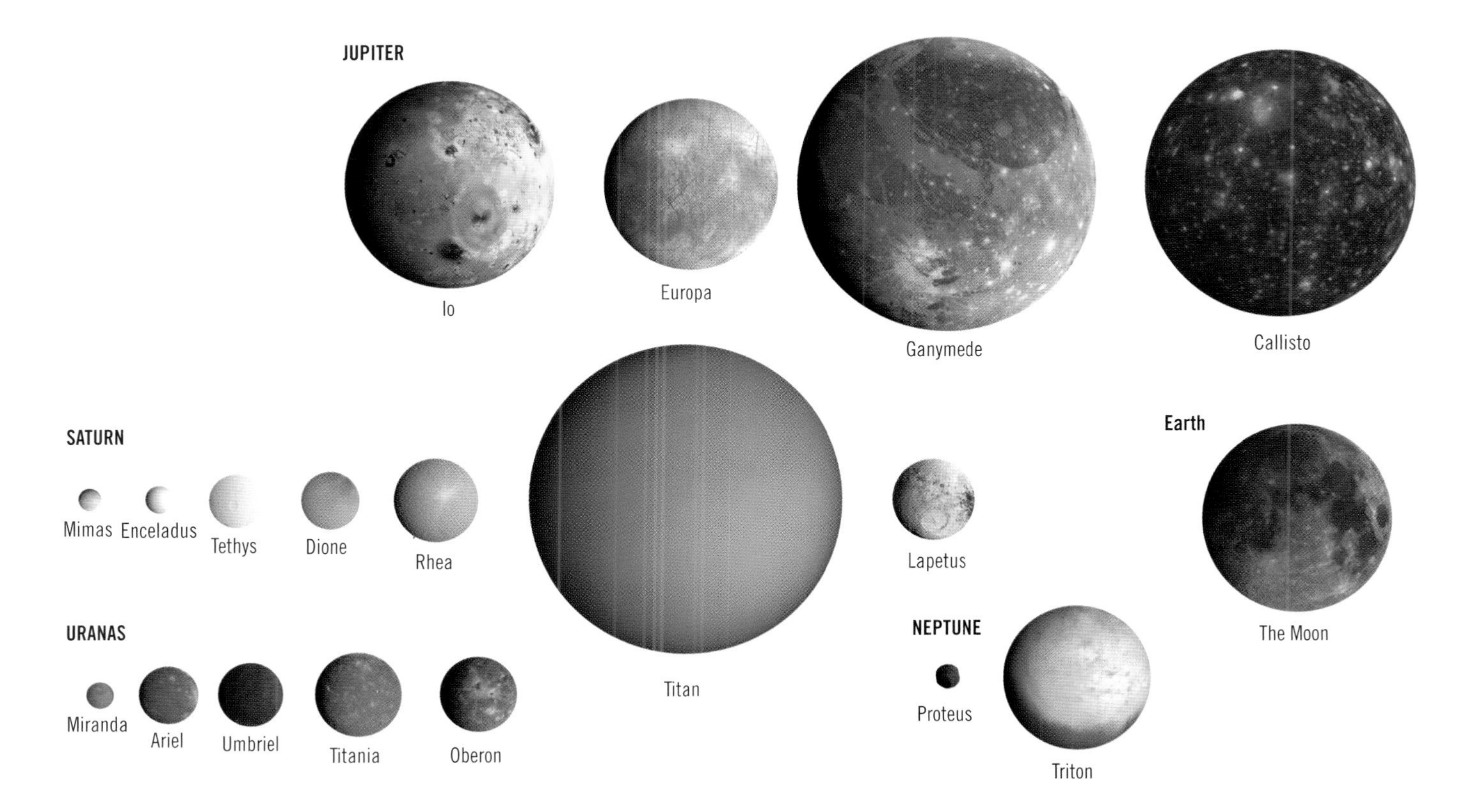

Characteristics of the solar system's major moons

Name	Primary	Diameter (miles)	Composition	Notes
Ganymede	Jupiter	3,273	Ice and rock	Sub-surface ocean
Titan	Saturn	3,199	Ice and rock	Only moon with an atmosphere
Callisto	Jupiter	2,995	Ice and rock	Oldest cratered surface
Io	Jupiter	2,263	Rock	Most geologically active body
Moon	Earth	2,159	Rock	Nearest astronomical body
Europa	Jupiter	1,939	Ice and rock	Sub-surface ocean
Triton	Neptune	1,681	Ice and rock	Captured Kuiper Belt object
Titania	Uranus	979	Ice and rock	Moon-like surface with ocean or "mare"
Rhea	Saturn	948	Ice	Heavily cratered surface
Oberon	Uranus	946	Ice and rock	Red and moon-like surface
Iapetus	Saturn	932	Ice	Large hemispheric brightness contrast
Charon	Pluto	753	Ice	Huge tidal rifts caused by Pluto

near passes of this moon beginning in 2004, followed by the Huygens lander arriving on its surface in 2005. The radar images could penetrate the atmosphere and map out numerous liquid basins and lakes along with river tributaries not unlike those of Earth. At a temperature barely over -290°F (-179°C) its surface is in a cryogenic deep freeze. Nevertheless, gasoline-like compounds such as ethane and propane take the place of liquid water on Earth, and Titan also has its own cryo-volcanoes that are hot enough to spew out liquid water as lava.

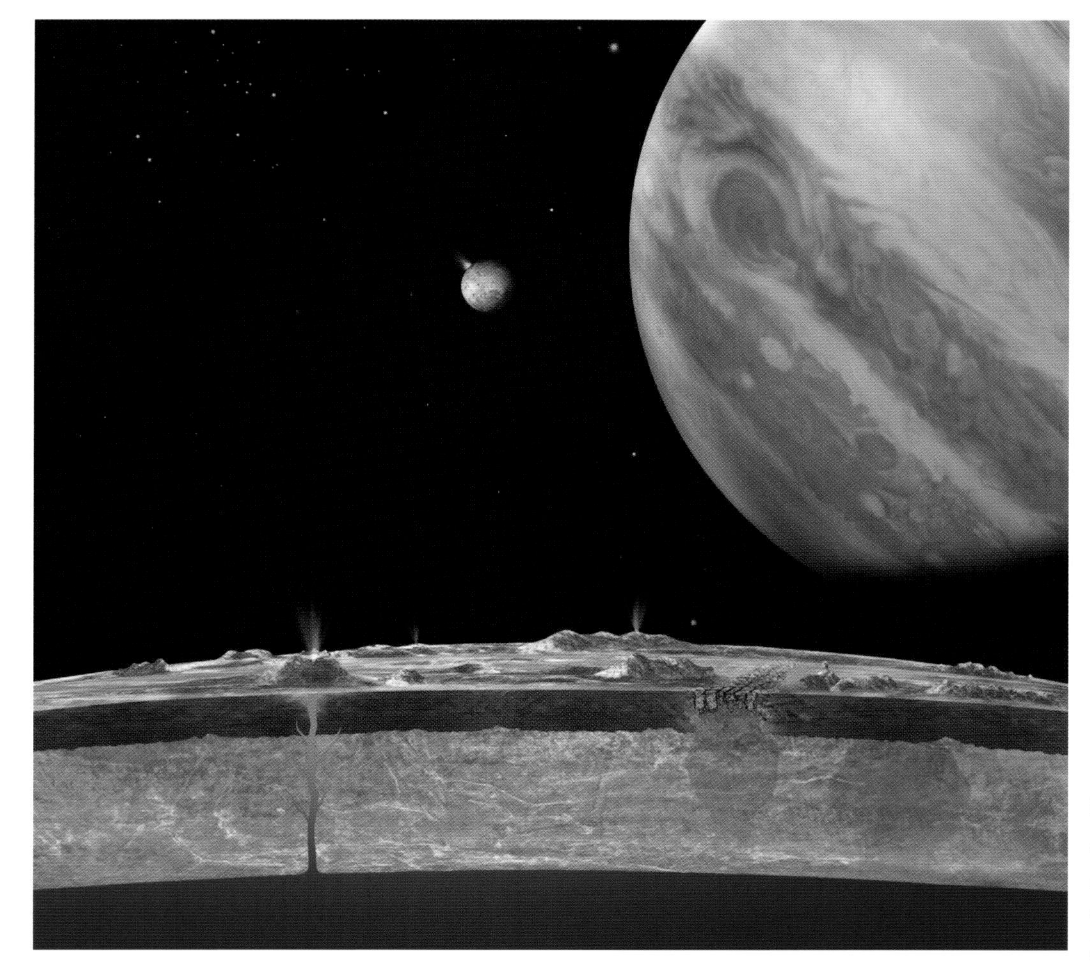

Below *The liquid oceans of Europa lie below a frozen crust of ice and may harbor life that emerged from its own chemical soup billions of years ago.*

Above *An artist's impression of a dust storm on Titan.*

Above *Apollo 15 astronaut Jim Irwin works in the Hadley-Apennine region of the moon during his mission in 1971.*

The moon

About one-sixth the size of Earth and orbiting in synchrony once every 28 days, our moon is gravitationally locked so only the same face ever reveals itself to Earth-bound observers. Despite its luminosity in the night sky, it actually has the reflectivity of worn asphalt. Its brighter, mountainous highlands are among the oldest surface rocks

Below *No photograph of the moon ever does it justice. This mosaic created by NASA's Lunar Reconnaissance Orbiter spacecraft gives a typical near-side view revealing its classic craters and dark mare.*

dating from the formation of the moon over 4.3 billion years ago. The dark mare or "oceans" are the magma outflows from enormous asteroidal impacts, which in some cases such as Mare Imbrium or Mare Serenitatis still show the circular shapes of vast impact craters filled-in by magma.

It remains the only astronomical body that humans have stepped foot upon, following the now 50-year-old Apollo program. Soon, another armada of visitors both robotic and human will return to its soil during the 2020s. From the Apollo visits by 12 astronauts, the returned soil and rock samples have largely confirmed how the moon was formed. About 100 million years after Earth formed, a Mars-sized planetoid collided with Earth and ejected vast quantities of surface material, together with the material in the planetoid. This mixed material took up residence around Earth as a ring of rubble, which over the millennia slowly accreted into the moon.

Phobos and Deimos

These moons of Mars are so small, barely 14 miles (22 kilometers) across, that they are more like asteroids than indigenous satellites formed by the accreting planet itself. The moons are so close to Mars, within 14,300 miles (23,000 kilometers), that their motion across the sky can be discerned from the surface of Mars. Both are in jeopardy of being eventually destroyed by tidal stresses and by orbit decay within the next few million years. Based upon their appearance and likely composition, they are probably captured asteroids, although the precise mechanism for how this "two-body" capture could have occurred is not understood. These moons may be the first destinations for Earth travelers because, with

Left *Phobos has a huge impact crater called Stickney (right edge) and linear features indicating strong gravitational stresses from its proximity to Mars.*

limited fuel, it is far easier to reach these moons from Earth than it is to complete the journey to the Martian surface at the bottom of a deep gravitational well. In fact, the masses for these moons are so low that travelers would actually dock with their surfaces rather than expend any rocket fuel at all.

Galilean moons of Jupiter

While moons are hard to find in the inner solar system, the outer gas and ice giants resemble miniature solar systems in their own right, with numerous satellites from three-feet-sized icebergs to Mercury-sized planetoids. The four largest moons of Jupiter were first spotted by Galileo and so are called the Galilean moons: Io, Europa, Ganymede, and Callisto. The proximity of Io to Jupiter results in the gravitational distortion of this world and a huge energy source that heats its interior. This results in over 400 volcanic vents spewing out compounds of sulfur that give it its orange-yellow color. Meanwhile, Europa, Ganymede, and Callisto generally have rocky cores overlain with internal oceans covered by a thick crust of ice. This ice crust is fractured by impact craters and the gravitational stresses caused by Jupiter. These are the closest objects in the solar system to Earth where life may be likely among the liquid oceans, and are primary targets of exploration for future robotic missions in the next few decades.

Below *The four largest "Galilean moons" of Jupiter along with their surface features imaged at medium- and high-resolution from spacecraft.*

Io **Europa** **Ganymede** **Callisto**

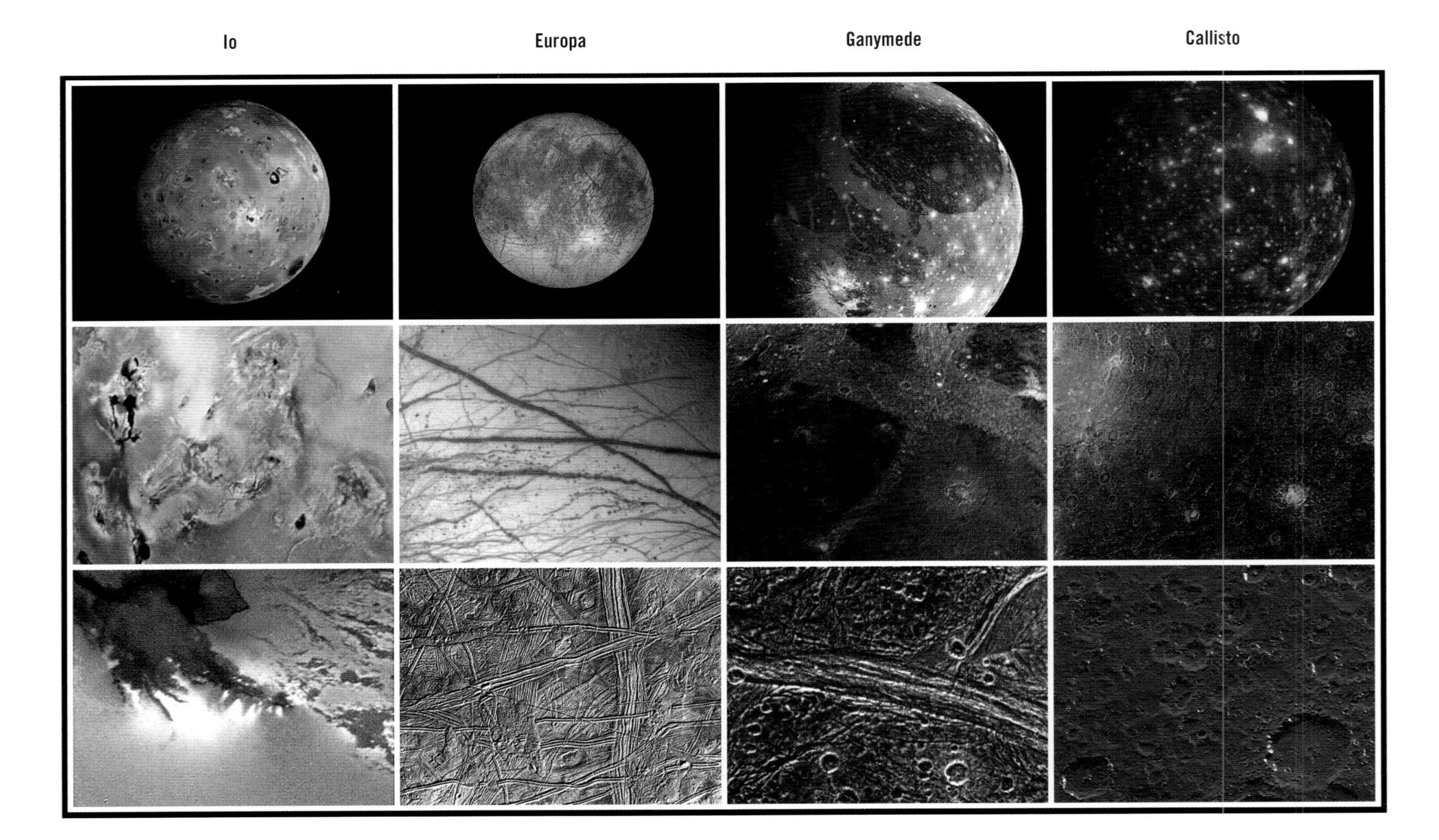

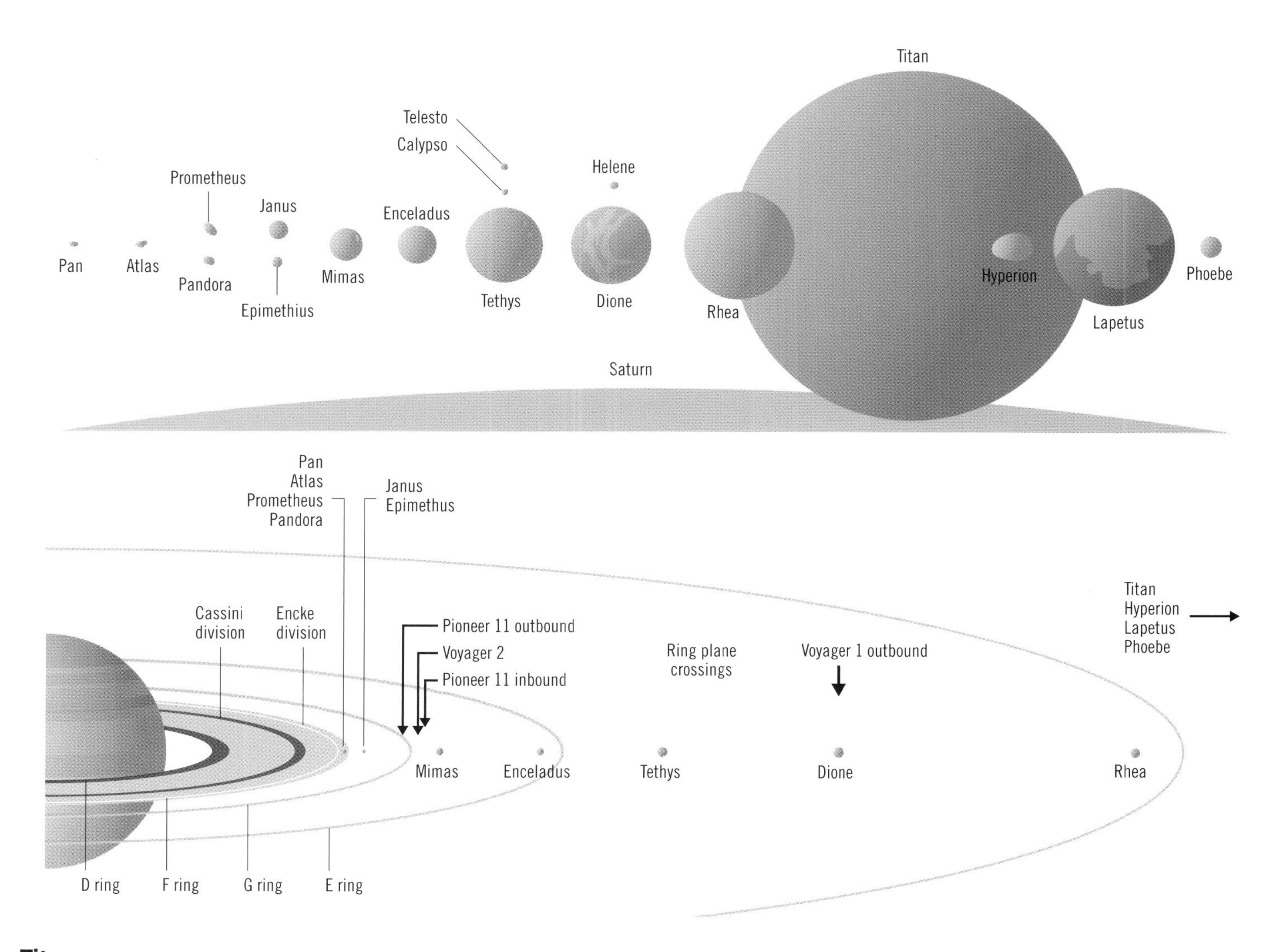

Titan

Among the dramatically colorful moons of Saturn, Titan is unquestionably the most enigmatic and exciting. Slightly larger than the planet Mercury, it has retained an atmosphere that is 50 percent as massive and dense as Earth's, although its composition is more in line with what you might find at a gasoline refinery. The Cassini/Huygens mission revolutionized our understanding of Titan through its surface lander Huygens and its imaging radar systems, which could cut through the atmosphere and map the surface below. The landscape is a bizarre version of Earth where ice takes the place of rock in mountains, river channels are cut by flowing methane and ethane, and lakes fill up with ethane over time. At a surface temperature of -292°F (-180°C) its organic-rich materials suggest pre-biotic Earth but frozen permanently into a perpetual deep-freeze. There may be a sub-surface liquid water ocean in which organic life could emerge. The presence of cryo-volcanoes on the surface that spew out liquid water lava seem to suggest a liquid ocean below the surface and enough internal energy to keep it fluid.

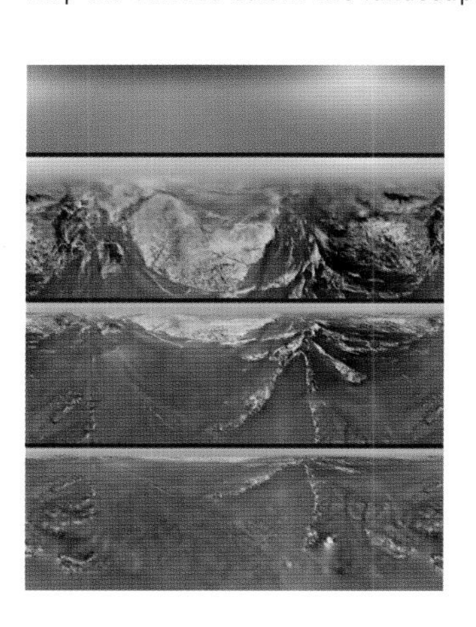

Left *Aerial views of Titan taken by the Huygens probe.*

Above *A comparison of the size and position of Saturn's moons.*

Below *The changing surface of Titan revealed in a series of infrared images taken by the Cassini spacecraft.*

Left *Uranus' icy moon Miranda is seen in this image from Voyager 2 taken on January 24, 1986.*

Right *It's believed that the rings of Neptune are relatively young—much younger than the age of the solar system, and much younger than the age of Uranus' rings. They were probably created when one of Neptune's inner moons got too close to the planet and was torn apart by gravity.*

Miranda

The largest moon of Uranus was a puzzle to astronomers when it was first imaged in high-resolution by the Voyager 2 spacecraft in 1986. Like most other large moons, its round shape suggested a normal formation by gravitational accretion, but its bizarre surface features suggested a more complex and less gradual formation. There are patchworks of material with numerous parallel grooves surrounded by cratered terrain that appears to be very old. According to one theory, Miranda may have been struck and fragmented by impact with a large body, then gravitationally re-formed from the pieces in a jumbled order. Another theory, more in-line with the detailed shapes and cratering of the landscape data, is that the interior of Miranda was at one time molten. Convection currents brought the internal material rich in water and ammonia to the surface through numerous surface fractures, where it solidified.

Triton

The largest moon of Neptune, and the seventh-largest moon in our solar system, is just slightly smaller than our own moon. At its distance from the sun, it receives only 1/900th the light that falls on our own moon and this keeps its surface at a temperature of -391°F (-235°C). It is very likely to have been a captured Kuiper Belt object due to its orbiting Neptune in the opposite direction of its other satellites. Its surface is covered by frozen nitrogen, but is geologically active with few impact craters and evidence of cryo-volcanoes that emit liquid water and nitrogen. The largest of these, at more than 62 miles (100 kilometers) across is Leviathan Patera. Its ejecta plumes extend over 5 miles (8 kilometers) into space and may provide opportunities for future spacecraft to pass through them and search for traces of organic chemistry and even life.

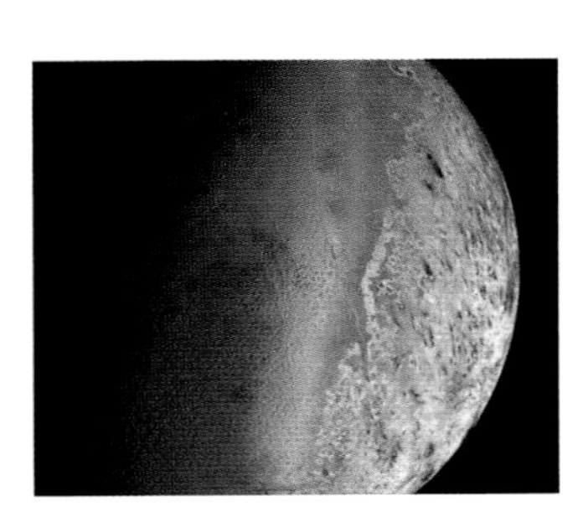

Above *Triton is the largest moon of Neptune, and also is the seventh-largest moon in the solar system. Geologically active, it has a surface of mostly frozen nitrogen, a mostly water ice crust, an icy mantle, and a substantial core of rock and metal accounting for two-thirds of its total mass.*

Charon

Charon orbits the dwarf planet Pluto, and is a satellite of considerable size and interest. No one expected that in these distant cold reaches of the solar system there would be much to see in terms of interesting surface features. However, Charon orbits Pluto once every six days and so it is subject to internal heating from gravitational, tidal stresses. The images returned from the New Horizons spacecraft reveal a complex surface torn by deep chasms and pelted by craters. Its variegated colors including deep crimson reds, suggest a landscape chemically processed by cosmic rays and dim solar ultraviolet light, but with some materials transferred from Pluto to Charon over the course of billions of years. Its interior is predicted to be a 50/50 mixture of rock and ice, with a rocky core and a mantle that is probably liquid water. Cryo-volcanoes appear on the surface, so Charon remains geologically active today. One of the most peculiar features is Kubrick Mons, also called the Mountain in the Moat. The 1.9-mile (3 kilometres) tall mountain may be an extinct cryo-volcano sitting in the middle of a frozen and depressed moat of water ice from the collapsed sub-surface chamber.

Below *NASA's New Horizons captured this high-resolution enhanced color view of Pluto's moon Charon just before its closest approach on July 14, 2015.*

// Asteroids

Above *The Asteroid Belt is situated between the orbits of Jupiter and Mars, but many thousands of asteroids can also be found within the orbits of the inner planets. Many of these are potential impact hazards for our Earth.*

Between the orbits of Mars and Jupiter, the majority of the solar system's rocky asteroids orbit the sun like mini-planets. They are the unconsolidated rubble left over from the protoplanetary disk out of which they condensed and grew through multiple collisions. They do not come in a fixed size, but vary from grains of sand to a handful of objects over 62 miles (100 kilometers) across. Although no spacecraft has as yet returned samples from them, meteorite fragments land on Earth and can be recovered for detailed study.

Asteroids come in several distinct families including stony (S-Type), iron/nickel (M-Type), and carbonaceous (C-Type). The stony asteroids orbit nearest the sun and are rich in silicate material resembling the surface of Earth's continents. The iron/nickel asteroids are almost pure iron and nickel and suggest that at one time there were several large asteroids whose interiors segregated the heavier materials into a core region, which after impacts now provides the parentage for iron/nickel meteorites. Finally, there are the carbonaceous asteroids rich in carbon compounds and organic materials including water. These asteroids have not been subjected to heat and are the pristine, unprocessed material that once dominated the protoplanetary disk. Also, within some meteorites one can see individual grains, some up to a few inches in size that probably originated in the atmospheres of ancient stars and enriched the protoplanetary disc with heavier elements.

Below *Asteroids and comets visited by spacecraft reveal a wide variety of cratered and rocky surfaces.*

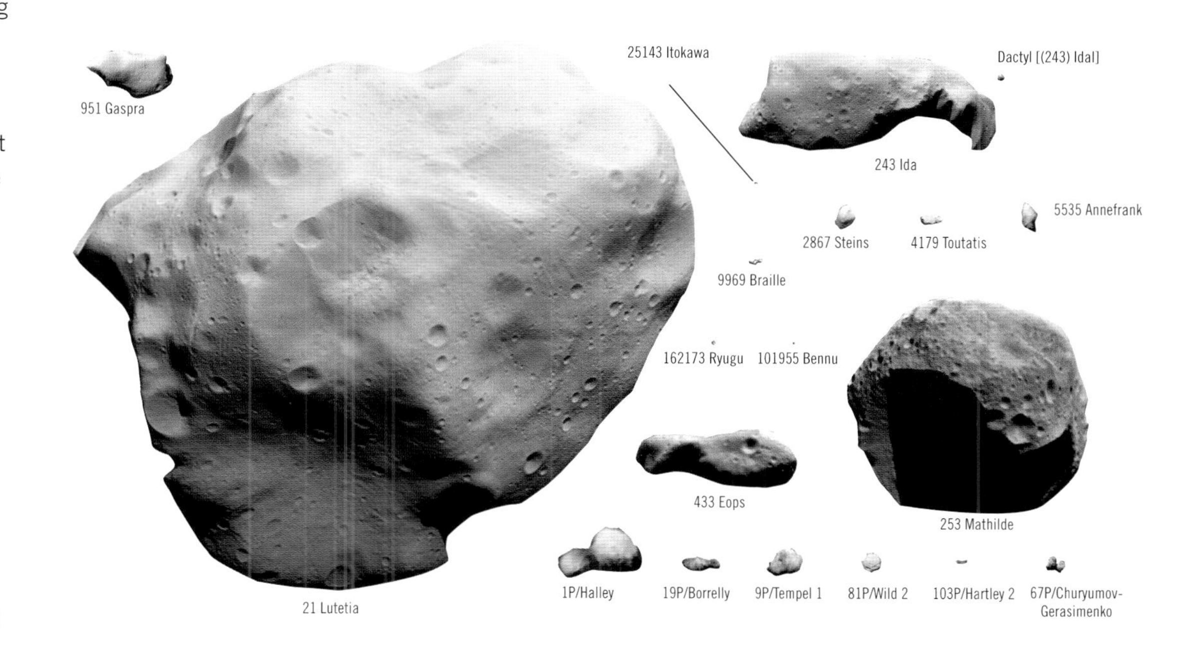

Above *This image shows the wide variety of boulder shapes, sizes and compositions found on asteroid Bennu. It was taken by NASA's OSIRIS-REx spacecraft from a distance of 2 miles (3.4 kilometers).*

Taken together, the total mass of the main asteroid belt is about 3 percent that of our moon, and represents a single object about 746 miles (1,200 kilometers) in diameter.

Although it is not possible to say exactly how many asteroids there are, one can estimate how many are larger than some particular size. For instance, astronomers have been able to discover virtually all of the asteroids larger than about 62 miles (100 kilometerss) and there are 240 of these. For asteroids larger than 0.62 miles (1 kilometer) however, there are estimated to be nearly two million. Astronomers are particularly interested in those larger than about 6.2 miles (10 kilometers) because just one of these that collided with Earth about 65 million years ago caused the extinction of the dinosaurs. Over 10,000 asteroids this large, or larger, have been discovered so far. Currently, the orbits of over one million asteroids have been determined. Spacecraft such as Dawn, Rosetta, Galileo, NEAR, Deep Space 1, OSIRIS-REx, Hayabusa and Chang'e 2 have also photographed their surfaces in detail. Most asteroid surfaces appear to be heavily cratered, although despite the numbers of asteroids in this region of space, collisions between asteroids are actually very rare events. In 2010, astronomers used the Hubble Space Telescope to study one of these events in the asteroid belt dubbed P/2010 A2. A larger body, perhaps 330 feet (100 meters) across, was impacted by a smaller asteroid. Although the larger body survived, the smaller one was reduced to a trail of rubble forming the tail.

Although the vast majority are located beyond the orbit of Mars, nearly 14,000 asteroids cross the orbit of Mars and enter the inner solar system. Astronomers are especially concerned about those that are within a few million miles of Earth's orbit, or cross Earth's orbit, because one of these could eventually impact Earth. Over 23,000 near-Earth objects, or NEOs, have been discovered to date and of these 2,093 are called potentially hazardous objects (PHOs) because their orbits are predicted to take them within 5 million miles (8 million kilometers) of Earth. About 157 of these are larger than 0.62 miles (1 kilometer) and so present a severe hazard and even an extinction-level event should one of them impact Earth. New NEOs and PHOs are discovered every year and are typically about 330 feet (100 meters) in diameter. Astronomers estimate that about 30 percent of these PHOs have been discovered so far, but their orbits have to be carefully updated every few weeks because of gradual changes due to gravitational encounters with Earth and the other planets.

Below *The comet-like object P/2010 A2 is probably a collision event between two objects that occurred about one year before this image was taken. The surviving asteroid is the star-like point seen at the far left.*

// Comets

Next to total solar eclipses, comets are among the most dramatic objects observed throughout recorded history. They consist of a large solid body up to 31 miles (50 kilometers) in diameter, composed of various ices, which begin to evaporate as the object approaches the sun. As the material vaporizes, it leaves behind a long tail of gas and dust that can stretch up to several million miles from the comet's nucleus or "coma."

Some comets, such as the famous Halley's Comet, return to the inner solar system on periodic, elliptical orbits. There are about 730 comets that have been cataloged so far. Between five to ten new ones are discovered every year. Most are so faint and distant they are not easily visible from Earth without the aid of a telescope. Returning comets with periods less than about 200 years are called the periodic comets. Their orbits tend to extend as far from the sun as the Kuiper Belt. Apparently, the constant gravitational jostling of these Kuiper Belt objects by the outer planets place some of them on elliptical paths that reach the inner solar system.

The second major class of comets are the long-period/non-periodic comets, whose orbits are either extremely elliptical or are hyperbolas. They have orbit periods considerably longer than 200 years, which makes observing their returns either impossible or highly unlikely. The most recent of these was Comet NEOWISE seen in 2020, which may not return for another 6,800 years. Their orbits seem to reach out into a distant reservoir of objects called the Oort cloud located some 10,000 AU from the sun. An example of one of the 2,500 of the known hyperbolic comets not likely to return

Below *After traveling more than ten years, ESA's Rosetta spacecraft reached comet 67P/Churyumov-Gerasimenko in 2014. This image was obtained from a distance of 177 miles (285 kilometers) above 67P's surface. This comet nucleus is dumb-bell shaped with a length of about 2.5 miles (4 kilometers). Visible are steep slopes and precipices, sharp-edged rock structures, prominent pits, and smooth, wide plains.*

Above *Comet NEOWISE was a dramatic naked-eye comet in July 2020 that was photographed numerous times by astronomers and the general public.*

to our solar system is Oumuamua, discovered in 2017. Its orbit trajectory reached escape velocity from the sun and is actually considered an interstellar object rather than a comet.

Spacecraft have visited several comets so far, beginning with the international Giotto spacecraft flyby of Halley's Comet in 1986 and most recently the Rosetta orbital encounter of Comet 67P/Churyumov-Gerasimenko in 2014. Although the Halley's Comet flyby was clouded by the dense cloud of ejecta caused by solar heating, the Rosetta mission returned thousands of high-resolution images of a craggy, boulder-strewn surface revealing a surprisingly complex landscape. Over time, some of the vents became active and the emission of gas into the forming cometary tail could be observed first-hand.

The origin of many cometary nuclei appears to be the Kuiper Belt, which is known to contain hundreds of ice-rich bodies several miles across. In 2019 the New Horizons spacecraft (launched in 2006) completed its flyby of one of these Kuiper Belt objects following its historic flyby of Pluto in 2015. The object had been identified as a target of opportunity and would be along the trajectory of the spacecraft after the Pluto encounter. Object 2014 MU69 was discovered in 2014 and officially named Arrokoth. It is only 22 miles (36 kilometers) across and orbits the sun every 297 years at a distance of 4 billion miles (6.7 billion kilometers). The spacecraft images showed that it consists of two smaller objects about 9 miles (15 kilometers) across that have fused together from an ancient collision. Spectroscopic studies of its surface show a reddish object with methanol, hydrogen cyanide, water ice, and other organic compounds. Like many distant objects that have a reddish color, these organics have been processed by billions of years of solar ultraviolet light and cosmic rays to become tholins, which have this characteristic color.

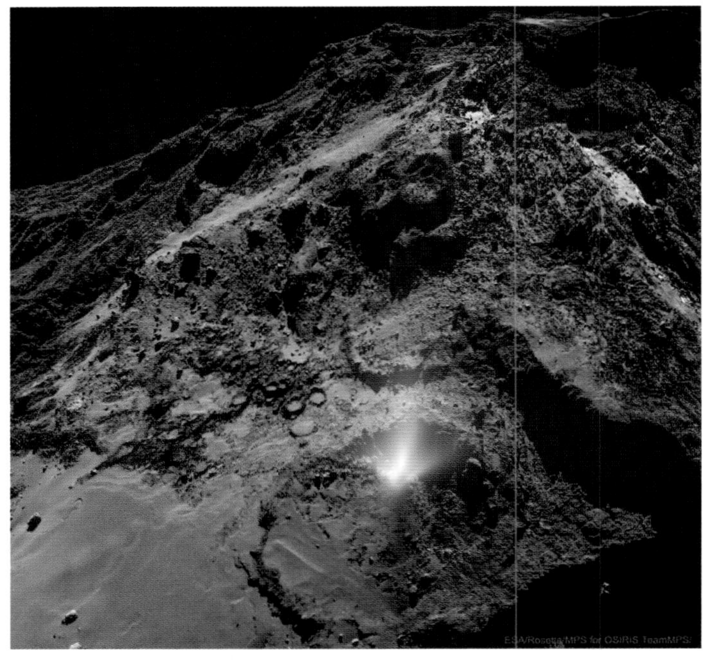

Above *A dust jet from the surface of Comet 67P.*

Below *The pristine Kuiper Belt object Arrokoth was studied by the New Horizons spacecraft. It is one of the furthest-known objects in our solar system to date whose surface has been imaged in detail.*

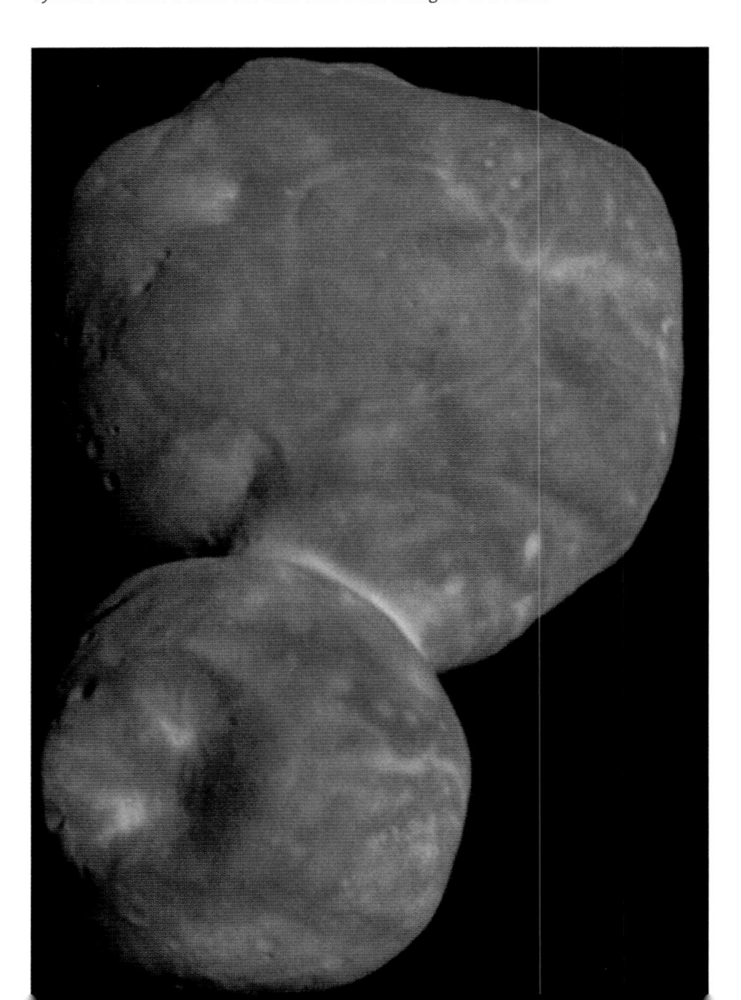

// Meteorite and asteroid hazards

We tend not to regard the meteors we see in the evening sky as a potential threat like getting struck by a lightning bolt, but in fact meteorites have been known to injure and even kill an unlucky few people over the centuries. Like asteroids, meteorites come in a range of sizes, from harmless sand grains that give us the numerous meteor streaks we see in the sky, to breath-taking fireballs caused by three-feet-sized objects that light up the entire sky and cast shadows on the ground.

Because meteorites surviving to the ground can be collected and studied, from them we can learn about the chemical composition of their parent asteroids. Rocky and stony meteorites have come from asteroids that were large enough for gravity to heat their interiors and cause the lighter silicate compounds to be separated from the heavier elements such as iron and nickel. After this differentiation period, these bodies were apparently smashed to bits by collisions. The left-over, silicate-rich and iron-rich fragments became the stony and iron/nickel meteorites we collect on Earth. The most pristine material that has never been subjected to heat and pressure are the carbonaceous chondrite meteorites. These are among the most intriguing samples from distant asteroidal bodies that are also among the oldest materials we can recover anywhere in the solar system. The Murchison meteorite fell in Australia in 1969 and had a total mass exceeding 220 pounds (100 kilograms). It is very rich in organic compounds, amino acids and 12 percent of its mass is water. Although its age is about 4.9 billion years, there are minute grains of silicon carbide that have ages approaching seven billion years. These grains were created in the atmospheres of long-extinct red giant stars and traveled across interstellar space to become embedded in the materials from which the Murchison meteorite accumulated.

Although the smaller meteorites burn up in the atmosphere, larger ones make it all the way to the ground. These can be recovered for study, or to be sold to collectors for thousands of dollars per kilogram, but problems arise when meteors are large enough on impact to leave behind craters. In remote areas, this is no problem or hazard, but if the crater is large enough to engulf a building, town, or city it is no longer a matter of idle entertainment.

Humans have been struck by small meteorites of masses less than a kilogram and fortunately many of these incidents have not been fatal. Cars and buildings have also been damaged but not devastated. It is a matter of probability and statistics how often large meteorites will impact Earth, with the largest 6.2 mile (10 kilometer) extinction-level asteroids happening every 100 million years or so, and the smaller metre-sized bodies encountering Earth every hour. In between these extremes, the size ranges lead to a

Below: *This fireball sighting map between 1988 and 2020 shows how common asteroid impacts are in the size range from about 3–230 feet (1–70 meters). The small circles indicate about 1,120 tons (1,016 tonnes) of TNT-equivalent energy release while the large red circle, representing the Chelyabinsk Impact of 2013, released 336,000 tons (305,000 tonnes) of TNT-equivalent energy. The smaller impacts occur about ten times a year while the larger ones occur once a century on average.*

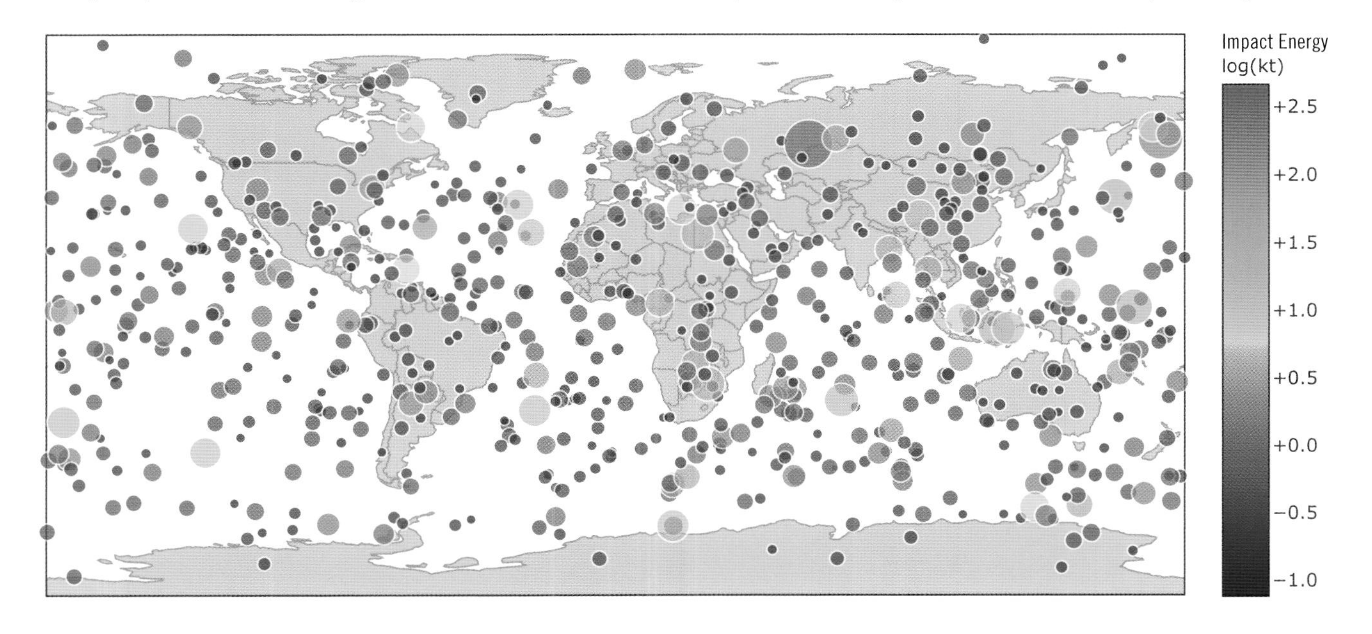

Above *The Murchison meteorite's pristine interior is rich in organic molecules but the surface is deeply scored and blackened from high-speed atmospheric entry.*

Left *A rare carbonaceous chondrite from a meteor fall in the Sahara desert. The numerous chondrules pre-date the formation of Earth and some may have been created in the atmospheres of long-dead red giant stars.*

decreasing frequency but an increasing level of devastation.

Astronomers have begun programs of detecting asteroids larger than 330 feet (100 meters) because, at this size, the impacts occur about once every century, but are capable of producing significant damage to entire cities. Two of these events have happened since 1900: the Tunguska Impact of 1908 and the Chelyabinsk meteor of 2013. Both of these bodies exploded at high altitudes before they impacted. The Tunguska meteor was probably 164–656 feet (50–200 meters) in diameter and delivered a 20 megaton TNT airburst that flattened an unpopulated Siberian forest. The Chelyabinsk meteorite in 2013 was about 66 feet (20 meters) in size with an airburst equivalent to 336,000 tons (305,000 tonnes) of TNT, nevertheless, it did significant damage to a small city, causing 3,000 injuries from flying glass and considerable building damage from the pressure wave.

The orbits of many NEOs have been calculated forward in time to determine whether they are actual collision hazards to Earth. Called potentially hazardous asteroids (PHAs), or potentially hazardous objects (PHOs) if comets are included, none of the best candidates have higher than 1-in-20 odds of impacting Earth, and are moreover less than 330 feet (100 meters) in size. Unfortunately, only 9,335 of the estimated 25,000 NEOs with sizes greater than 459 feet (140 meters) have been found to date so there are still plenty of undiscovered NEOs and PHAs yet to be worried about. The list of PHAs changes daily as new ones are discovered and old ones have their orbits improved from subsequent measurements. What is disturbing is the time between their discovery and when they are estimated to become a hazard can be less than a few years in some cases. This is not enough time to launch any intercept missions but enough time for evacuation planning once the likely target area on Earth can be calculated. Given the uncertainties, however, the eventual target zone may be as large as the entire hemisphere of Earth's surface, which is not helpful.

Below: *The Chelyabinsk meteor entered Earth's atmosphere on February 15, 2013. Its airburst explosion damaged buildings and caused injuries.*

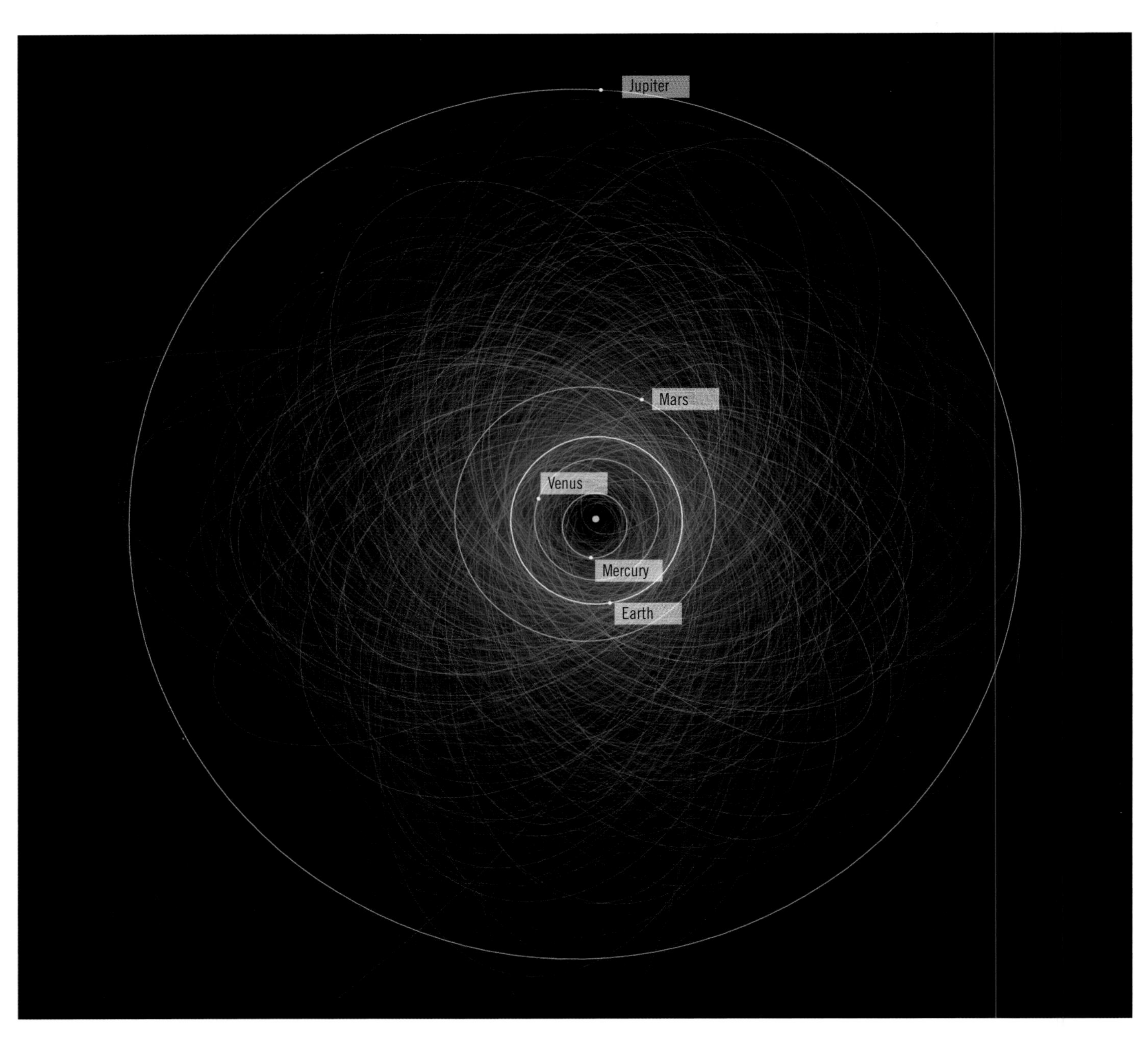

Name	Size (ft)	Closest Approach	Date
99948 Apophis	1,066	38,300	April 13, 2029
2007 UW1	245–560	100,200	October 19, 2129
2012 UE34	165–395	107,800	April 8, 2041
2007 YV56	395–1,215	235,200	January 2, 2101
2005 YU55	1,050–1,310	237,000	November 8, 2075
2000 WO107	1,215–2,755	243,700	December 1, 2140
2011 WL2	625–1,380	244,600	October 26, 2087
2001 WN5	2,300–4,925	248,800	June 26, 2028
1998 OX4	560–1,215	296,200	January 22, 2148
2005 WY55	625–820	332,500	May 28, 2065
2009 DO111	245–490	335,200	March 23, 2146

Above *Graphic showing the orbits of all the known potentially hazardous asteroids (PHAs), numbering over 1,400 as of early 2013. These are the asteroids considered hazardous because they are fairly large (at least 459 feet/140 meters in size), and because they follow orbits that pass close to Earth's orbit (within 4.7 million miles/7.5 million kilometers).*

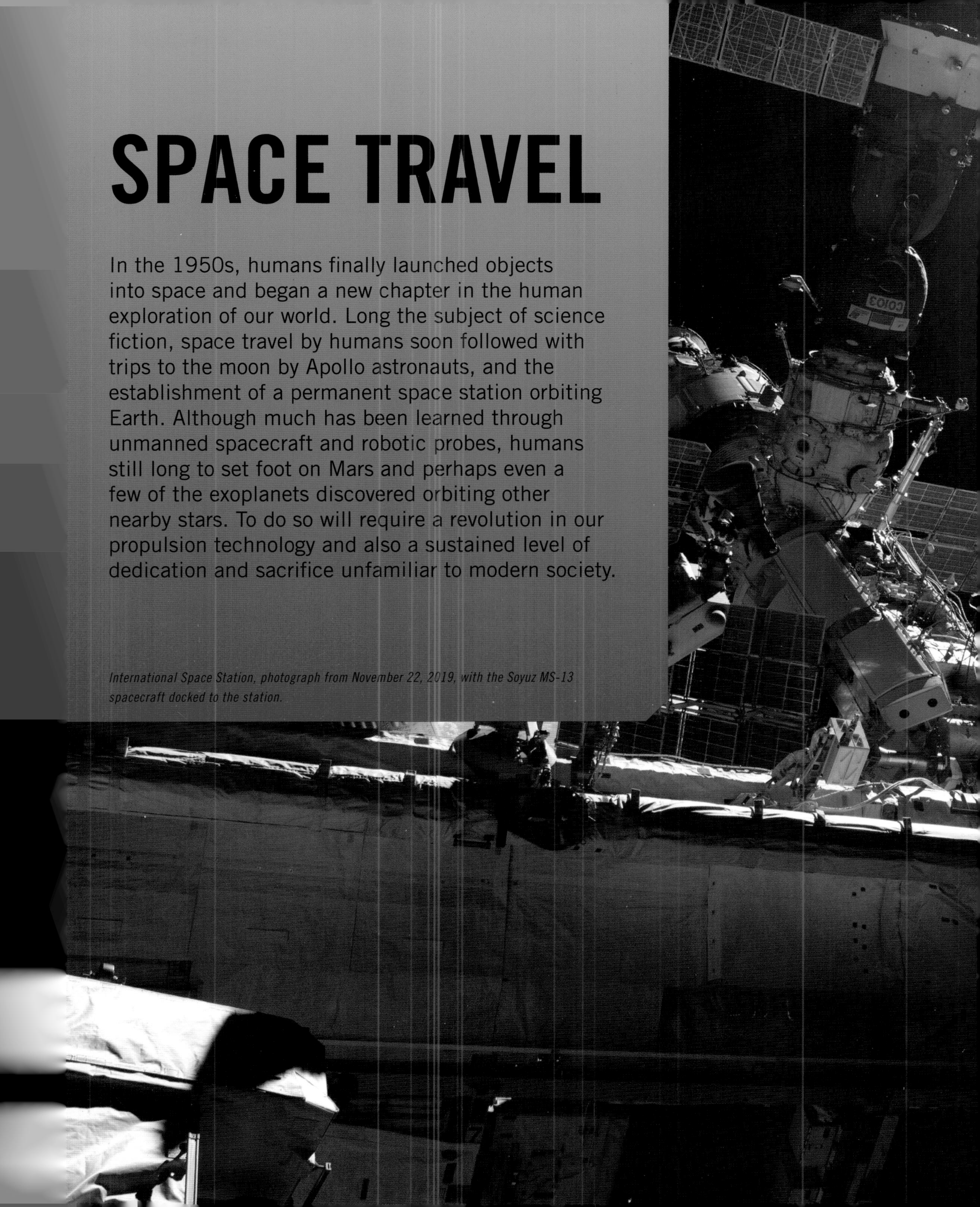

SPACE TRAVEL

In the 1950s, humans finally launched objects into space and began a new chapter in the human exploration of our world. Long the subject of science fiction, space travel by humans soon followed with trips to the moon by Apollo astronauts, and the establishment of a permanent space station orbiting Earth. Although much has been learned through unmanned spacecraft and robotic probes, humans still long to set foot on Mars and perhaps even a few of the exoplanets discovered orbiting other nearby stars. To do so will require a revolution in our propulsion technology and also a sustained level of dedication and sacrifice unfamiliar to modern society.

International Space Station, photograph from November 22, 2019, with the Soyuz MS-13 spacecraft docked to the station.

// Remote adventures

Humans are natural explorers, largely because millions of years ago as hunter-gatherers it was necessary for us to follow the animals we subsisted upon. This urge to travel over the millennia has led us to explore the entire surface of Earth. Only in recent centuries have we gazed up at the moon and stars and convinced ourselves that, at some point in the future, they too will become a new frontier for us to explore. Space exploration began in earnest in the 1950s with the launching of a few primitive satellites such as Sputnik and Explorer 1 to great political fanfare. Within a fast-paced 60 years, we have walked upon the moon, and sent robotic spacecraft to Mars. Our spacecraft, following a decade or more of travel, have even reached the outer limits to our solar system far beyond Pluto. It has become an international adventure with over a dozen countries having their own space program. Chief among these efforts has been unmanned exploration.

For centuries, we have been conducting unmanned exploration of the universe through remote sensing using powerful ground-based telescopes to gather data. With the advent of rocketry capable of carrying heavy payloads into Earth orbit and beyond, it is now possible to explore our solar system using remote sensors, cameras, and robotic systems. The information gathered during flybys or orbits of planets and asteroids is then transmitted back to Earth for analysis. Some of the most spectacular information has been provided by the Hubble Space Telescope, launched in 1990, which has returned over one million images of planetary surfaces, stars, and distant galaxies and nebulae. Spacecraft such as COBE, WMAP, and Planck have revolutionized our understanding of the origin and evolution of the universe. Spacecraft such as the Voyager 1 and 2, Cassini, Galileo, New Horizons, and Juno provided the first-ever images and measurements of the environments of the outer planets. Mars has been heavily studied by dozens of flyby, orbiter, and lander missions, including the spectacularly successful, Curiosity rover launched in 2012. The Mars Reconnaissance Orbiter, launched in 2005, over the course of 45,000 orbits, has returned over 65,000 high-resolution images. Closer to home, the entire surface of our moon has been mapped to 3.3 feet (1 meter) resolution by the Lunar Reconnaissance Orbiter launched in 2009. Future unmanned missions include a mission to land on Venus, missions to orbit and study the larger moons of Jupiter for signs of liquid water, and larger and more capable Mars landers and rovers to search for signs of fossil life. The low costs of unmanned exploration, rarely more than a few billion dollars, along with their enormous scientific returns, make a case for this effort to continue indefinitely into the future with only our imaginations to limit the questions we will ask.

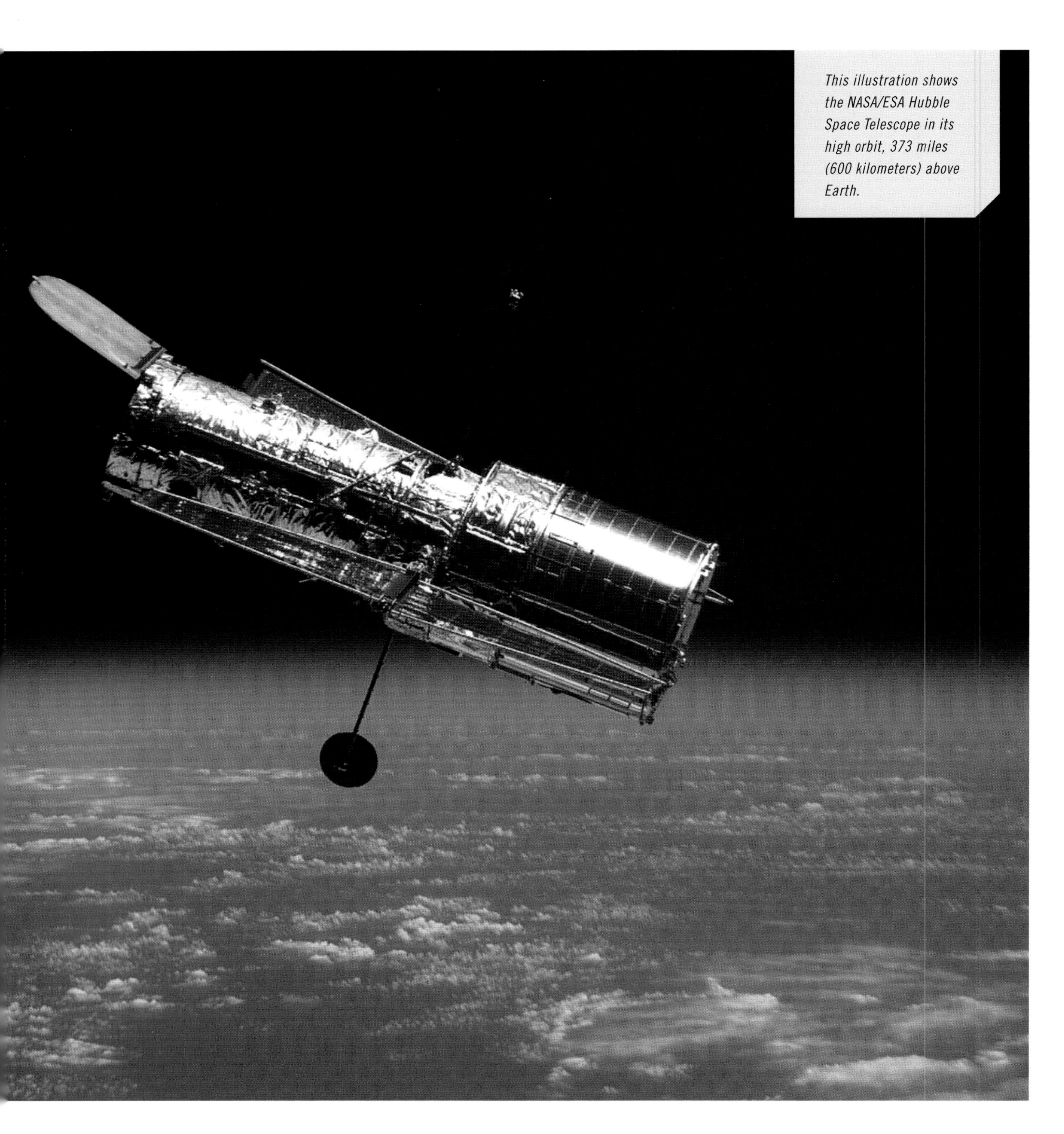

This illustration shows the NASA/ESA Hubble Space Telescope in its high orbit, 373 miles (600 kilometers) above Earth.

// Manned exploration

The exploration of Earth began almost as soon as humans appeared and benefitted from a number of free resources: there was always air to breathe at the proper pressure; gravity always promised the same direction and magnitude to keep our physiology and biochemistry working properly; the atmosphere shielded us from harmful ultraviolet and cosmic radiation that would otherwise damage our genes; there was always abundant liquid water available; and there was almost always plenty of food and shelter available to us wherever our journeys took us. We focused on the excitement and profit of exploration. There didn't seem to be any spots on our planet's land surface that we couldn't explore.

Once we leave our planet, however, none of the free resources of our nurturing biosphere are available. We have to bring them with us or create them using our technological cleverness. But to even start the adventure, we have to pay a steep price; we have to leave our planet's surface at the bottom of a deep gravity well. This requires building rockets that deliver huge thrusts very quickly. Typical journeys to orbit take less than ten minutes and require an energy expenditure of 5.9 million Joules per pound (13 million Joules per kilogram). That works out to about 10,000 watts per pound (22,000 watts per kilogram). The most compact and low-mass fuel we have uses chemical reactions with a fuel like liquid hydrogen and an oxidizer

Below *The International Space Station photographed by Expedition 56 crew members from a Soyuz spacecraft after undocking.*

Above *The astronaut Buzz Aldrin deploys a seismic sensor on the moon.*

like liquid oxygen. They explode on contact in the rocket chamber to deliver the thrust in ejected gases. Chemical rockets have been the backbone of space exploration for decades, resulting in progressively larger rocket systems as our goals have steadily increased. Manned exploration requires a breathable atmosphere, temperature control, and provisions that include food and water. Even short journeys to the moon take three days and require a rocket like the 31,361-ton (28,450 tonnes) Saturn V to leave Earth with its 134-ton (122 tonnes) human cargo and life support systems. During these short trips lasting only a week, there is little time for the human body to get into serious trouble. The astronauts return to Earth with no ill effects. Long-term stays on board the International Space Station (ISS) and planned trips to Mars are an entirely different matter.

Studies of astronauts living on board the Mir and International Space Station for up to 14 months have documented many harmful effects on humans of prolonged stays in space. Bone loss, vision impairment, and even genetic damage are among the more troubling effects. Progressive declines in the human immune system have also been detected. None of these impacts to human physiology have simple interventions to either eliminate or reduce their consequences, which increase as more days in space are logged. Journeys to the moon and back lasting a few months may not be severely affected, but typical journeys to Mars require up to 224 days in transit each way, plus a potentially long-term 1.5 year stay on the planet's surface, which although it has some gravity is poorly shielded from solar and cosmic radiation. Although strong arguments can be made that setting up research colonies on the moon are within the realm of human endurance, trips to other planets in our solar system are, for now, still at the level of simply an interesting exercise in engineering, and a potentially deadly one for human biology.

Below *The bulkhead wall thickness of aluminum provides some shielding from cosmic and solar radiation but not enough for missions lasting more than a few years. This image shows astronaut Reid Weisman on the ISS.*

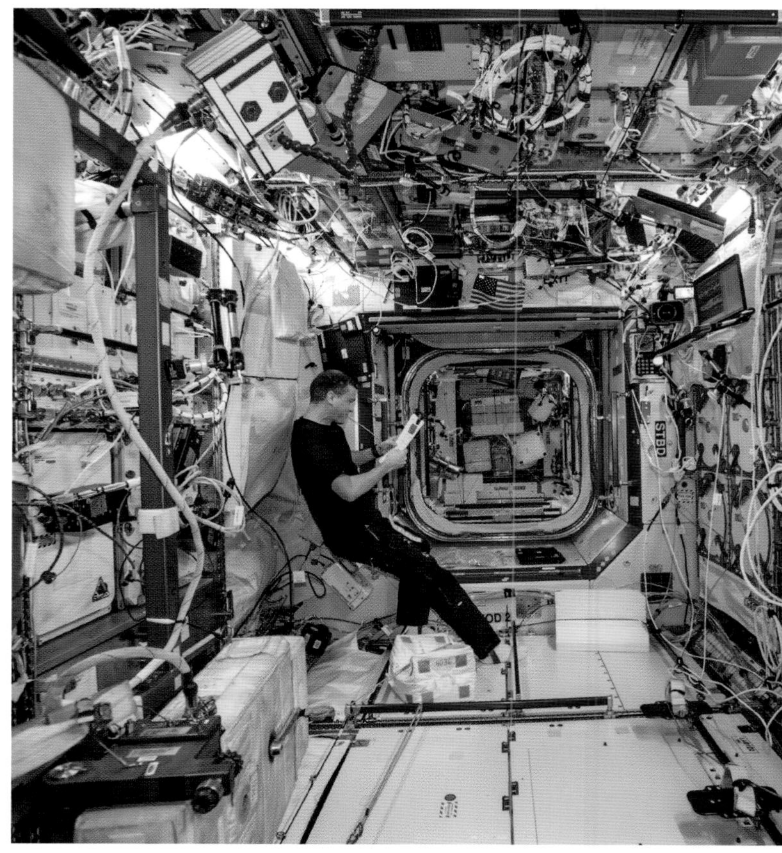

// Interstellar travel

The idea of traveling to the distant stars in search of liveable planets to colonize has been the stuff of science fiction since before the turn of the 20th century. Stories abound about routine interstellar travel to colony worlds, sprawling galactic empires, and contact with other civilizations. In some instances, rockets with a handful of explorers simply traveled to a nearby star to search for habitable worlds; finding none, they then hopscotched to other stars. In the real world, this is not how we would undertake such exploration at all given the multi-trillion-dollar cost of the spacecraft and the likely hundreds of years such a journey might entail. Just as for the exploration trips of our ancestors, there are three questions we would ask ourselves even before a mission had traveled its first mile: 1) What is our destination?; 2) What will we do when we get there?; and 3) What will be the value to the billions of people on Earth?

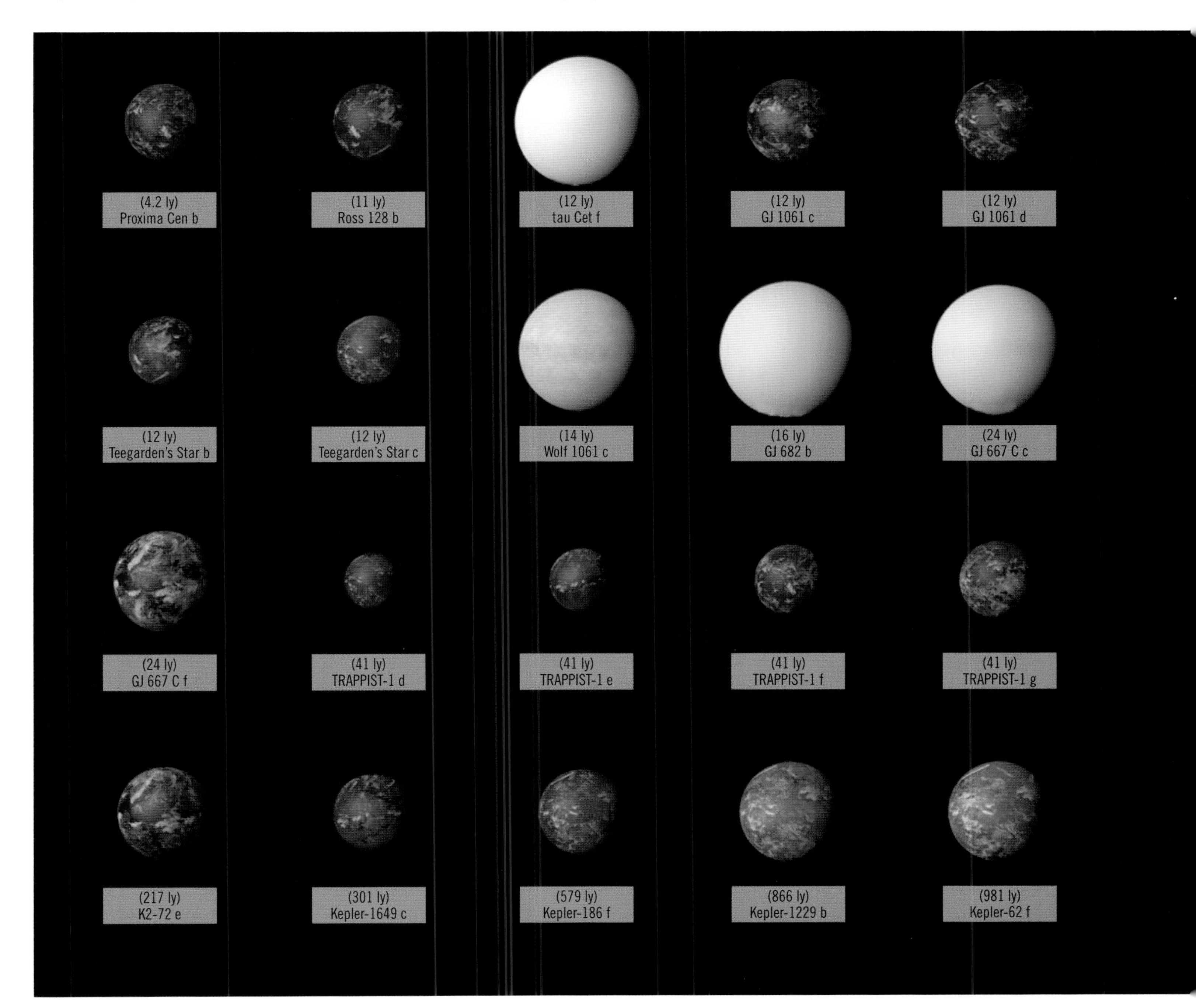

they could marshal 100,000 people to work for 20 years or more toward a religious goal that they were all emotionally invested in. But interstellar travel will require a human civilization that may look nothing like what we have today that could marshal such an expensive undertaking with no obvious benefit to the humanity left behind.

Although interstellar travel is a captivating idea for science fiction, fewer stories are told about the early years of this adventure when rocket technology was still being perfected and our furthest destinations were generally inside the asteroid belt. The means to solving the problem of traveling rapidly within our solar system is connected to the still-more-rapid speeds needed for interstellar travel. If you can travel from Earth to Pluto in a week, this is fast enough to eliminate all of the harmful effects upon humans of prolonged space travel and allow us to quickly set up colonies everywhere. But at this speed, it would still take you 130 years to travel from Earth to Alpha Centauri. Rapid interplanetary and interstellar travel can only be accomplished by moving beyond chemical propulsion into new systems. In science fiction, there is no end to the number of magical systems available to do this including faster-than-light "warp drives" and even networks of worm holes, but in the real world of engineering and physics, progress in rocket design is much slower. Rocket propulsion relies on throwing as much mass as you can out the back of your rocket in the least amount of time. Chemical rockets do this by ejecting plumes of gas at rates of many tons per second. It's a cheap, brute force way to get the job done. But since the 1960s there are other engine designs far more efficient and requiring far less fuel.

// Propulsion methods

Nuclear propulsion

This method, developed and prototyped in the early 1960s, involves setting up a nuclear fission reactor through which liquid hydrogen is quickly heated to thousands of degrees. While chemical rockets eject matter at a few miles per second, nuclear rockets can eject mass at up to 62 miles per second (100 kilometers per second), providing enormously higher thrusts for much less fuel. Unfortunately, designs such as the Phoebus 2A that was tested in 1968 do not produce enough thrust to carry their own mass into space, but they do work incredibly well once they are in space. Nuclear engines would first be delivered into orbit perhaps with a conventional chemical rocket. They would then be used as part of an incredibly rapid, interplanetary transport system that could shorten trips to Mars from eight months to perhaps less than a week.

In 2021, the United States Defence Advanced Research Projects Agency (DARPA) provided funds for three major companies, General Atomics, Blue Origin, and Lockheed Martin to develop nuclear thermal propulsion systems for operating within the orbit of our moon by 2025. The same year, NASA was also given $110 million to accelerate the design of these systems for future manned space travel to Mars in 2039.

Ion propulsion

A charged particle can be accelerated to speeds of thousands of miles per second by using a combination of electric and magnetic fields to accelerate and steer the particle. This basic idea of an ion engine was first considered by Robert Goddard in a prototype engine that operated in 1924, but it took another 50 years to apply ion propulsion to spacecraft. For commercial satellites such as PAS-5 and Galaxy 8i launched in the 1990s, it was an efficient means for maintaining satellite positions in the orbits.

The first interplanetary mission to use ion propulsion was Deep Space 1 launched in 1998. Its mission was to flyby the asteroid 9969 Braille and the comet 19P/Borrelly. The spacecraft used a 2,100-watt engine that ejected a stream of xenon atoms to produce a thrust of 0.092 Newtons, but it could maintain this thrust for 1.3 years and consume only 181 pounds (82 kilograms) of xenon fuel, but the exit speed of the ions was 19 miles per second (30 kilometers per second). The

Below *NASA awarded an $18.8 million contract to BWXT Nuclear Energy in August 2017. A spacecraft using this primitive version of nuclear-thermal propulsion could cut the travel time to Mars by 20 or 25 percent compared to chemical rocket engines.*

electricity was provided by solar panels so that this method is properly called solar-electric ion propulsion. This works well for travel in the inner solar system where sunlight is intense, but in the distant solar system a nuclear reactor would be used in a nuclear-electric ion propulsion system.

Currently, ion propulsion systems capable of providing continuous thrusts of 5.4 Newtons have been demonstrated at the University of Michigan under contract to NASA. Operating at a maximum electric power of 100 kilowatts, the X3 system, which weighs 500 pounds (227 kilograms), uses krypton or xenon atoms and consumes about 331 pounds (150 kilograms) of fuel in a typical 100-hour operation time. Other ion engine designs such as the Variable Specific Impulse Magnetoplasma Rocket (VASIMR) developed by Ad Astra, with an exhaust velocity of 31 miles per second (50 kilometers per second) operating at 120 kilowatts, are being seriously looked at for transporting astronauts to Mars. The round-trip travel time would be reduced from 2.5 years to only five months. This could be shorted to a 39-day trip by using a nuclear powered 200-megawatt reactor. Another more immediate application would be to use five, 1 megawatt VASIMR engines powered by solar panels and 8.9 tons (8.1 tonnes) of xenon propellant to transport 38 tons (34.5 tonnes) of cargo to the moon.

Above *NASA's new X3 thruster, which is being developed by researchers at the University of Michigan in collaboration with the agency and the US Air Force, has broken records in recent tests.*

Below *An artist's impression of several VASIMR engines propelling a craft through space.*

// Epilogue

In the span of a few hundred pages and a few hours of reading, you have created an entire universe—at least in your mind. The last hundred years of exploration have revealed how the many different ingredients to our world have had to come together in just the right ways to fashion the planets, stars, and galaxies that fill the night sky. Our modern story of the origin and evolution of our universe, meanwhile, has taken many unexpected twists and turns. We have had to form new understandings of time, space, and matter, leaving behind the millennia-old biases that have hampered creating better and more accurate stories. Nevertheless, ancient stories were sufficient for what our ancestors needed from them when the most advanced technology consisted of wheels and levers. Today, our technology-enhanced senses are vastly improved from those of our predecessors and so the sophistication of our storytelling must increase too.

Some may worry that the deep technical details of the new story are too remote from human experience to satisfy our emotional yearnings for simple explanations. But simplicity is a matter of taste. It is a moving target that changes constantly through the centuries. The biggest change to our story compared to ancient ones is that there are so many more parts to it, and there is a far-deeper thread of continuity that knits all of the parts together. We have discovered a much larger and richer universe of galaxies, stars, and planets, but one in which kitchen table laws of nature organize this orchestra of matter like an unseen conductor. Amazingly, with the modern re-discovery of our universe, it is still possible to tell our new story while sitting in front of a campfire. This time we can see through the clutter of foreground matter and phenomena and contemplate the deeper mystery of how it all came to be.

Below *Abell 78, a "born again" star.*

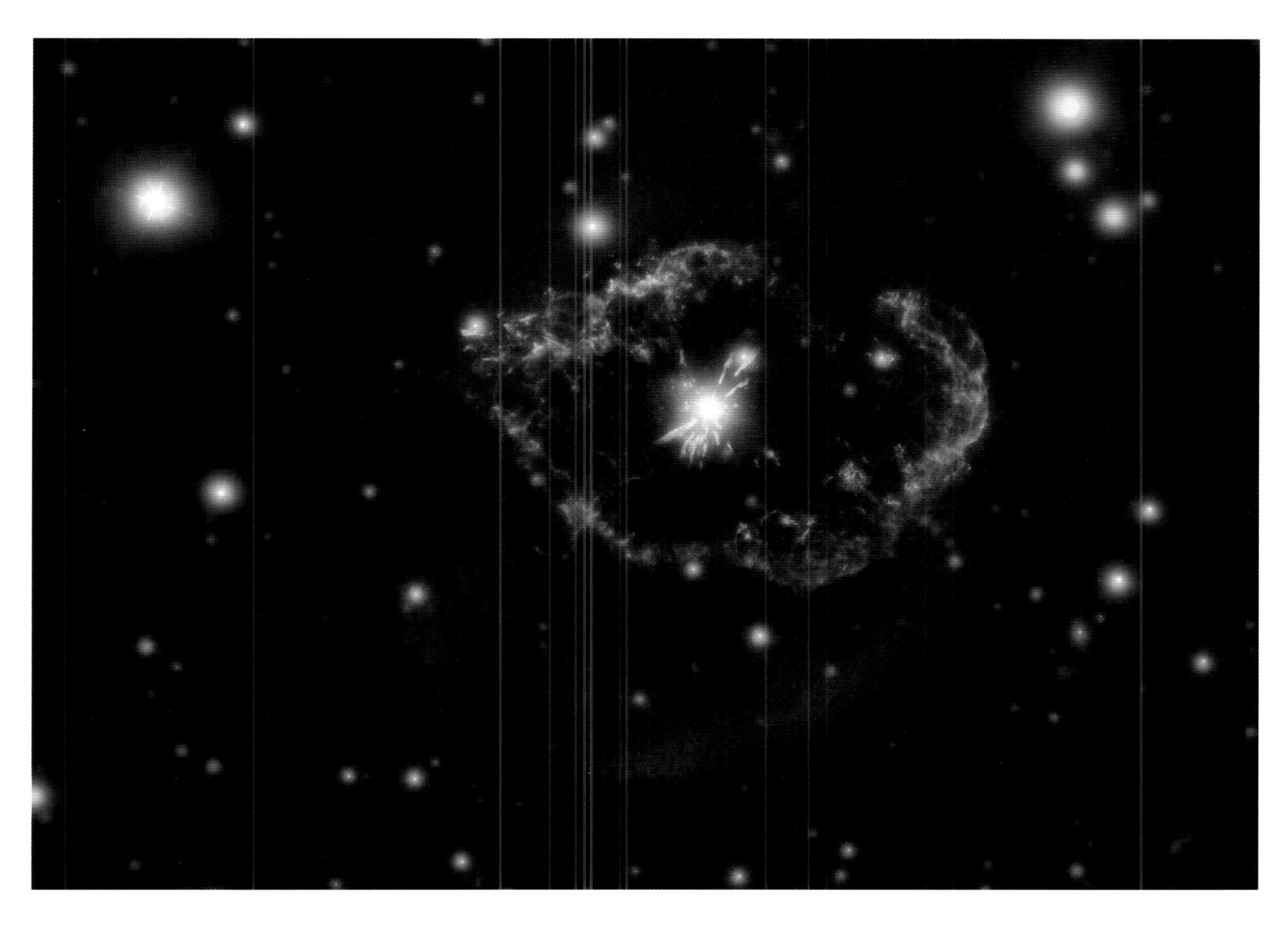

Further reading

The Meaning of Relativity by Albert Einstein, Princeton University Press, 1922.

Dark Side of the Universe: Dark Matter, Dark Energy, and the Fate of the Cosmos by Iain Nicolson, Johns Hopkins University Press, 2007.

The Theory of Almost Everything: The Standard Model, the Unsung Triumph of Modern Physics by Robert Oerter, Pearson Education Press, 2006.

Before the Beginning: Our Universe and Others by Martin Rees, Helix Books, 1997

Three Roads to Quantum Gravity: A New Understanding of Space, Time and the Universe by Lee Smolin, Basic Books, 2001.

A Brief History of Time by Stephen Hawking, Bantam Books, 1988.

Exploring Quantum Space by Sten Odenwald, CreateSpace Publishing, 2015.

Observer's Guide to Stellar Evolution by Mike Inglis, Springer, 2003.

The Life and Death of Stars by Kenneth Lang, Cambridge University Press, 2013.

How Did the First Stars and Galaxies Form? by Abraham Loeb, Princeton University Press, 2010.

Visual Galaxy: The Ultimate Guide to the Milky Way and Beyond by Clifford Hadfield, National Geographic, 2019.

Exoplanets: Diamond Worlds, Super Earths, Pulsar Planets, and the New Search for Life beyond Our Solar System by Michael Summers and James Trefil, Smithsonian Books, 2018.

Envisioning Exoplanets: Searching for Life in the Galaxy by Michael Carroll, Smithsonian Books, 2020.

Our Solar System: An Exploration of Planets, Moons, Asteroids, and Other Mysteries of Space, by Lisa Reichley, Rockridge Press, 2020.

The Planets: Photographs from the Archives of NASA by Nirmala Nataraj, Chronicle Books, 2017.

Interstellar Travel: An Astronomers Guide by Sten Odenwald, CreateSpace Publishing, 2015.

The Future of Humanity: Terraforming Mars, Interstellar Travel, Immortality, and Our Destiny Beyond Earth by Michio Kaku, Doubleday, 2018.

Index

Picture Credits

t = top, b = bottom, l = left, r = right, m = middle

Alamy: 37 (Science Photo Library), 173b (Susan E. Degginger)

Alex Alishevskikh: 174

Andrea Ghez: 101tl (Keck Observatory/UCLA Galactic Center Group)

CalTech: 81 (Virgo Collaboration/LIGO)

CERN: 48

Daniel Pomarede: 103t (Yehuda Hoffman/Hebrew University of Jerusalem)

David Woodroffe: 20, 35, 132, 154b, 166t

Derek Leinweber: 32 (CSSM, University of Adelaide)

Event Horizon Telescope: 95

ESA: 7, 15, 53, 67, 70t, 70b, 72–3t, 91t, 91b, 96, 112, 115b, 120l, 120r, 123t, 135, 138b, 157b, 164tl, 166bl, 170, 171tr, 176, 179, 188

ESO: 69, 82, 85, 86, 87, 90, 100, 106, 116, 117t, 121, 127t, 130

Gabe Clark: 109t

Getty Images: 11t (Print Collector/Hulton Archive), 19b (Mark Garlick/Science Photo Library), 79 (Mark Garlick/Science Photo Library)

Ken Chen: 73br (National Astronomical Observatory of Japan)

Kevin Jardine: 119b

Jan Skowron: 119t (OGLE/Astronomical Observatory, University of Warsaw)

NASA: 2, 4, 6, 8, 16, 17, 18, 21b, 22, 30, 43, 54, 59, 60, 62, 63, 65, 66, 74, 75tl, 75tr, 75b, 77, 89, 91b, 93, 94, 97bl, 97br, 98l, 98r, 99l, 99m, 99r, 101tr, 101b, 105, 109b, 114, 115t, 117b, 122, 123b, 126, 127b, 128, 131, 134, 139, 140, 142, 144, 145t, 145b, 146t, 146b, 147t, 149t, 150, 151t, 151b, 152t, 152b, 153b, 154t, 155t, 155b, 156, 157t, 158, 160l, 160r, 161, 163, 164tr, 164b, 165t, 165b, 166br, 167tl, 167bl, 167tr, 167br, 168t, 169t, 169b, 171b, 172, 175, 180, 181t, 181b, 184, 186, 187t

NRAO: 97tl (AUI/NSF/B. Saxton)

Philip Moesta: 73bl (TAPIR/California Institute of Technology)

PHL@UPR Arecibo: 133, 136, 182

Physical Review Letters: 45 (N. Musoke et al.)

Planetary Society: 162, 168b

Public Domain: 110

Science Photo Library: 13 (Arscimed), 23t (Take 27 Ltd), 28 (Mark Garlick), 34 (Giroscience), 41 (Mark Garlick), 47 (Laguna Design), 80 (Nicolle R. Fuller), 104 (Mark Garlick), 118b (Royal Astronomical Society), 136 (Mark Garlick), 143 (Tim Brown), 147b (Sputnik)

Shutterstock: 12, 19t, 27, 38, 44, 49, 55, 56, 71t, 124, 148b

TNG Collaboration: 103b

Wikimedia Commons: 10, 25, 58, 61, 64, 68, 102, 108, 111t, 113, 118t, 137, 153t, 159, 171tl, 173t, 187b